In Situ PCR and Related Technology

Jiang Gu
Editor

Soft Cover Edition, Eaton Publishing Co.
Hard Cover Edition, Birkhäuser Boston
1995

Jiang Gu
Deborah Research Institute
One Trenton Road
Browns Mills, NJ 08015

Library of Congress Cataloging-in-Publication Data

In situ polymerase chain reaction and related technology / edited by
Jiang Gu.
p. cm.
Includes bibliographical references.
ISBN 1-881-299-02-3 (Eaton Pub. : pbk. : alk. paper). -- ISBN
0-8176-3870-9 (Birkhäuser Boston : hbk. : alk. paper)
1. Polymerase chain reaction. 2. In situ hybridization I. Gu,
Jiang.
QP606.D46I52 1995
574.87'3282--dc20 95-25215
CIP

ISBN 1-881-299-02-3 (Soft Cover Edition, Eaton Publishing Co., 154 E. Central St., Natick, MA 01760, USA)

ISBN 0-8176-3870-9; and 3-7643-3870-9 (Hard Cover Edition, Birkhäuser Boston, 675 Massachusetts Ave., Cambridge, MA 02139, USA)

Printed in the United States of America

9 8 7 6 5 4 3 2

CONTENTS

The cover photo shows a double staining of human condyloma with in situ *PCR for human papillomavirus (HPV) as detected with the indirect IGSS method with silver acetate autometallography (black; infected nuclei are labeled) and immunohistochemistry for cytokeratins (S-ABC method using DAB as the chromogen; brown; epithelial cells are labeled). The tissue was fixed in formalin, embedded in paraffin and cut into 7-μm-thick sections. The preparation was lightly counterstained with hematoxylin. Control sections were not stained for HPV when* in situ *hybridization alone was used without prior* in situ *PCR amplification, or when primer or polymerase was omitted.*

Contributed by Gerhard W. Hacker (Salzburg, Austria), Ingeborg Zehbe (Uppsala, Sweden) and Cornelia Hauser-Kronberger (Salzburg, Austria).

Preface

Ever since the introduction of the polymerase chain reaction (PCR) in 1986, morphologists, whose interests lie in the analysis of intact tissue structures, have been attempting to adapt this technique to intact cells or tissue sections to detect low copy numbers of DNA or RNA *in situ* while preserving tissue morphology. The significance of this objective is obvious. A technique finally materialized in 1990 when Dr. Ashley T. Haase and coworkers published results that used multiple primers with complementary tails in intact cells. Since then, a number of laboratories have successfully developed their own versions of the technique. *In situ* PCR is now a well-recognized method that permits the detection of minute quantities of DNA or RNA in intact cells or tissue sections. As a result, morphological analysis of those target nucleotide sequences becomes possible. As anticipated, this advancement has led to significant improvement in our understanding of many normal and abnormal conditions, and its impact is becoming more evident as time passes.

In situ PCR has the characteristics of a new landmark in morphologic technology—it is scientifically fascinating and technically challenging. In essence, it is a combination of *in situ* hybridization and conventional PCR. The wealth of literature, experience and protocols for the two latter techniques can be applied to *in situ* PCR. *In situ* PCR also has its own unique aspects that were not addressed by the other two techniques. For example, *in situ* PCR requires that the target nucleotide be available for amplification without interruption, yet the templates have to be effectively anchored *in situ* by fixation and tissue processing. These can only be achieved by optimal fixation, tissue processing and enzyme digestion. Another critical step is the inhibition or reduction of amplified signal diffusion. In theory, it seems that there is no reason for the amplified products to remain *in situ* rather than to float in the supernatant of the reaction mixture. The fact is that PCR products, or at least some of them, do remain at the site of origin under certain conditions. While the subsequent detection or visualization of the amplified signals should not be an obstacle, the elimination of unspecific amplification binding or distinction of specific and unspecific reactions could be a formidable task. More controls are also needed for *in situ* PCR than for the other techniques. It is the search for these and other empirical conditions that makes the setting up of an *in situ* PCR protocol challenging. Those technical steps are so demanding that only a dozen or so laboratories have reported success in performing *in situ* PCR. This technique is at an exciting stage of development, and more simplified, straightforward versions and widespread applications are bound to appear. This book is intended to accelerate this evolution.

Thus far, applications of this technique have already broken ground in a number of disciplines, particularly in viral infections, tumor diagnosis, AIDS research and gene therapy. For those who want to detect minute quantities of DNA or RNA against a background of tissue structure, *in situ* PCR may be the only technique available. A number of other techniques, such as fluorescence *in situ* hybridization and some enzyme-driven reactions have similar promise; however, they all have inherent limitations. *In situ* PCR is branching into a number of variations, each with its particular attractive features. Some can be combined with immunohistochemistry, image analysis, confocal microscopy or electron microscopy, thereby broadening the horizon for morphologists and many other professionals.

These chapters were written by experts in the field of *in situ* PCR and related techniques. The authors have had considerable experience in establishing and performing the procedures. Their experience at the bench is particularly valuable for those who intend to perform these techniques. This book is organized to give theoretical consideration, historical review and practical, step-by-step guidance in the protocols for performing *in situ* PCR and its variations. It is hoped that with the help of this book, readers will gain a balanced view of this emerging technology.

I am indebted to the contributing authors who have shared their expertise and time to make this book possible. I am also grateful to Ms. Christine McAndrews for her excellent copyediting work and to Dr. Tak-Shun Choi for his invaluable input. It is our hope that this volume will make a valuable contribution to the literature and to those who intend to use these techniques in their research or practice.

Jiang Gu
August 14, 1995

In Situ PCR—An Overview

Jiang Gu

Deborah Research Institute, Browns Mills, New Jersey, USA

SUMMARY

In situ *polymerase chain reaction (PCR) is a morphological technique used to detect minute quantities of DNA or RNA in tissue sections or intact cells to unveil their distribution. It is derived from a combination of the conventional PCR widely employed by molecular biologists and* in situ *hybridization commonly used by morphologists. The former is capable of amplifying minute quantities of DNA or RNA in test tubes to billions of identical copies for analysis, but it does not allow the correlation of amplified signals with tissue structure. The latter can visualize DNA or RNA sequences in tissue samples and correlate them with tissue structure, but it has limited detecting sensitivity and requires relatively large amounts of DNA or RNA for detection. A combination of the two enables one to pinpoint up to a single copy of DNA or RNA of interest in tissue sections or intact cells and allows subsequent identification of cellular structures. This capability gives new insight into the cellular distribution of low copy numbers of nucleic acid sequences in many situations, such as viral infections, gene mutations, gene alterations, chromosomal translocation, gene therapy and low level gene expression. It is opening up a number of new territories for exploration and is expected to have a significant impact on basic and clinical research and diagnosis. This chapter gives a comprehensive account of the theoretical and practical background of this technology, presents its major variations and considers the commonly used protocols step-by-step. Its current and future applications in research and diagnosis are also discussed.*

INTRODUCTION

Since first reported in 1990 (26), *in situ* PCR technology has undergone rapid development. Many modifications and variations have been made and a number of applications have been reported. Owing to its unprecedented detecting sensitivity, some applications have led to important discoveries in several disciplines. In common with any new technology at its incipient stage, controversy about its reproducibility and significance of results is unavoidable. For the most part, these discussions are healthy and serve to propel the improvement of the method. The technology is still very young, and the procedures are quite complicated and technically demanding. Continuing evolution of the whole technique or parts thereof is inevitable. Its impact on a number of fields, such as viral infection, gene mutation, genetic disorders, gene therapy and low level gene expression, is expected to be significant.

Conventional PCR

In order to understand *in situ* PCR, one must understand conventional PCR. Since its invention by Mullis et al. in 1986 (42), PCR has had a profound impact on molecular biology. It has provided an extremely sensitive and relatively straightforward means to amplify very small amounts of DNA or RNA—down to a single copy of a gene, to milligram amounts of the same sequences consisting of millions or billions of identical copies for detection, sequencing, cloning, diagnosis, etc. PCR was popularized by the introductions of automatic thermocyclers and thermostable DNA polymerase—*Taq* DNA polymerase. Before its inception, DNA amplification could only be achieved by time-consuming cloning that took several days. PCR can be set up and completed in a matter of hours. The importance of PCR is exemplified by the rapid increase in the number of publications relating to PCR from three in 1986 to more than 1700 in 1990. In 1993, Mullis shared a Nobel prize for this discovery. The principle of PCR is illustrated in Figure 1 (C–H).

DNA is a double-stranded nucleic acid chain consisting of two complementary nucleic acid strands made up by four basic nucleotides—dATP (A), dCTP (C), dGTP (G) and dTTP (T). The four nucleotides are linked to one another by phosphodiester linkage. The two strands are bound together by hydrogen bonds in a complementary fashion with A bound to T with two hydrogen bonds and C bound to G with three hydrogen bonds. The annealing and separation of the two strands depend on a number of factors, particularly temperature, salt concentration, pH, nucleotide composition and length of the sequence concerned. The denaturation or "melting" temperature (T_m) is the point at which 50% of the double-stranded DNA is separated. It can be calculated according to the following formula (6,7,9):

$$T_m = -16.6 \log (Na) + 0.41 (\%GC) + 81.5°C - 0.61 (\% \text{ formamide}) - (500/\# \text{ base pairs in DNA/DNA hybrid})$$

where T_m = melting temperature, (Na) = cationic concentration and %CG = percentage of CG nucleotide. Formamide concentration is irrelevant to PCR but is important in hybridization and washing. The optimal annealing temperature is 25°C below T_m. For example, for a pair of nucleotide strands containing 50% of G and C, in a PCR buffer at 50 mM KCl, T_m will be about 80.4°C, and the best annealing temperature will be 55.4°C. This formula is more applicable to long strands of DNA rather than to the short primers. However, it demonstrates the relationship among the different factors in PCR reaction.

DNA polymerase is an enzyme that can make a complementary strand of a single-stranded DNA. The heat-resistant enzyme, *Taq* DNA polymerase can synthesize DNA at a theoretical rate of about 150 nucleotides per second per enzyme molecule at around 75°–80°C, although the polymerization step in most PCR protocols is set at around 72°–75°C (13,18). It is important to note that *Taq* DNA polymerase has a built-in error rate of about 0.1%–0.3% (60). Other DNA polymerases, such as *Vent* and *Pfu*, may have proofreading ability but amplify at a lower efficiency (41). The polymerization reaction takes place in the presence of DNA polymerase, magnesium and the four free nucleotides. Single-stranded specific nucleic acid sequences, named primers, usually 18–28 nucleic acids in length, serve to initiate specific binding sites for the making of complementary strands. The free nucleotides are added one-by-one by the DNA polymerase, according to the complementary template to one end of the annealing primer. The newly synthesized strand elongates from the 5′ to 3′ direction. For PCR, a pair of such primers is used, each

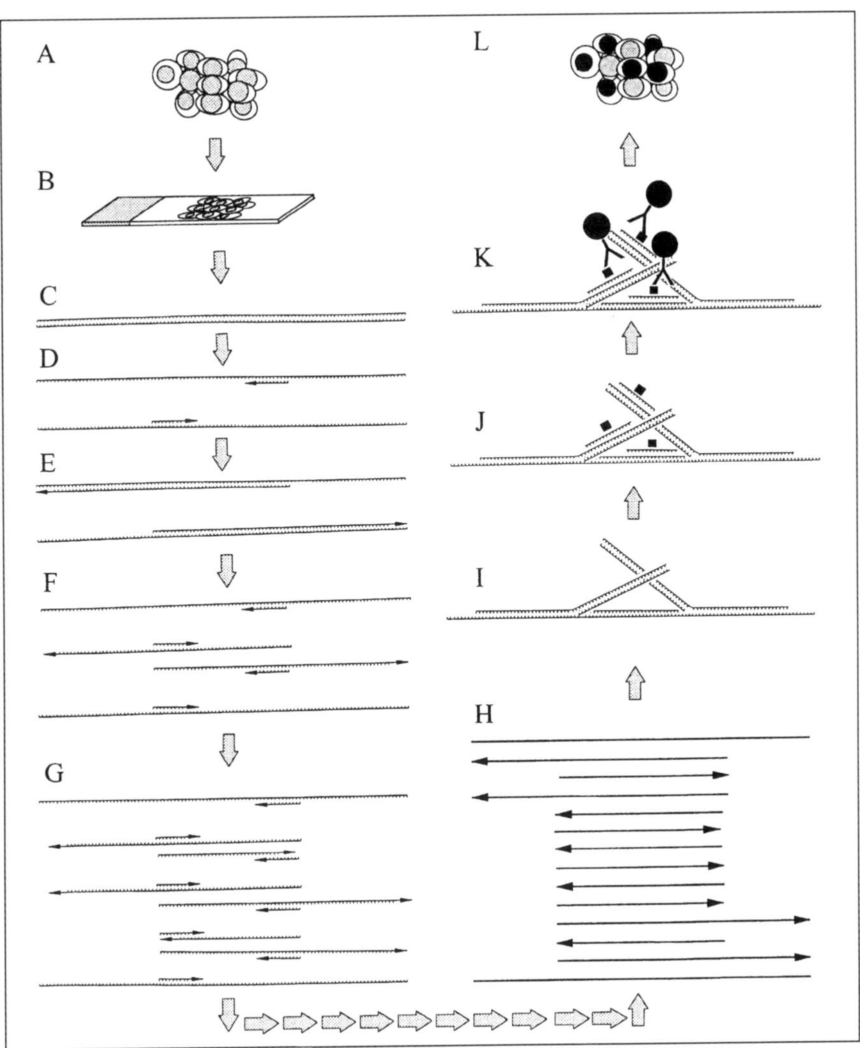

Figure 1. Diagrammatic illustration of the major steps of indirect *in situ* PCR. A) The samples can be individual cells in suspension or solid tissue samples. B) In most cases, the samples can be made into cytospin, smear or tissue sections. Chromosomal preparations may also be used. For cell suspension, *in situ* PCR may be performed in test tubes and then spun down on slides for detection. C) DNA in the nuclei is double stranded before denaturation. D) Above 90°C, DNA is denatured and becomes single-stranded DNAs that are bound by a pair of primers at specific sites. E) The primers serve as the starting sites for initiation of the polymeration reaction and form two new complementary chains of nucleotides. These two chains extend beyond the boundary defined by the primer on the opposite strand of the DNA template. F) The second cycle repeats the first cycle and uses the original DNA as well as the two newly synthesized strands as templates for making new complementary strands. G) The third cycle repeats the process to make more products. H) The reaction repeats itself in a thermocycler. By the end of n cycles, the sequence encased by the 5′ of the two primers (short products) becomes 2^{n-1} copies and the long products become $2 \times n$ copies. I) A possible mechanism for the amplified products to stay *in situ* is that the long products serve to anchor some of the short products as well as to form networks among themselves at the sites of the original templates. J) The amplified sequences and the original sequences are detected by specific probes which are labeled by, for example, digoxigenin. K) Digoxigenin is recognized by specific antibodies and visualized by a variety of markers. L) The cells that contain the very small amounts of target DNAs are thereby amplified and detected on the background of cellular morphology.

complementary to a specific sequence on one strand of DNA. The two primers are usually 100–1000 bases apart. This duplication reaction occurs at an optimal temperature range depending on the reaction condition. Beyond this range, the reaction still takes place but at lower rates (13). PCR is performed in a thermocycler that can be programmed to increase and decrease temperature from about 0°C to 100°C at about 1°C or more per second and maintain at a certain temperature for defined durations. The PCR typically consists of three different temperature gradients. First, at about 94°C, the double-stranded DNA are denatured and become single-stranded. At about 45°–65°C, the primers anneal to the specific complementary sequences of the denatured single-stranded DNA. At about 72°C, the DNA polymerase will add nucleotides onto the complementary strands and complete the extension. After each cycle, the sequence between the 5′ of the two primers doubles. The two newly synthesized strands serve as additional templates for subsequent PCR reactions. The sequence encased by the two primers grows in number exponentially thereafter following each cycle. Theoretically, after 30 cycles, a single copy of DNA sequence will become 10^9 identical molecules. For one target of 250 bp fragments, for example, this will represent approximately 0.3 ng of DNA, easily detectable by Southern blot analysis (13). The above reaction can also be used to amplify single-stranded RNA by first reversely transcribing the RNA into cDNA with a reverse transcriptase followed by the same PCR (30). Once PCR is completed, the end product can be detected by electrophoresis. The bands can be identified by size or by hybridization with labeled nucleotide probes that recognize a specific sequence of interest. A variety of labeling techniques ranging from radioisotopes to proteins can be used to label the probes (12,27). A number of controls are needed. These include the use of negative or positive samples, omission of primers, pretreatment of the samples with DNase or RNase, etc. (13). They all serve to ensure the specificity and the sensitivity of the amplification. Despite its extremely high sensitivity, conventional PCR cannot tell in which cells a particular DNA or RNA sequence is found.

In Situ Hybridization

In situ hybridization entails a hybridization reaction between a labeled nucleotide probe and a complementary strand of target DNA or RNA in tissue sections or intact cells. The hybridization reaction follows the same principle of nucleic acid chain annealing and disassociating as for PCR. Depending on what labeling methods are used, there are several means to visualize the reaction. The principle of *in situ* hybridization is illustrated in Figure 1 (J and K).

During the past decade, *in situ* hybridization has gone through many refinements and has become one of the most important tools in detecting DNA and RNA sequences in tissue samples. In 1992 and 1993, for example, more than 3000 articles utilizing *in situ* hybridization were published. The appropriate conditions for various factors in *in situ* hybridization have been established. These factors include tissue fixation, pretreatment, prehybridization, temperature, incubation duration, salt concentration, formamide concentration, probe length, concentration and composition, washing conditions and detecting systems (66).

Many variations of *in situ* hybridization have been developed. The probe can be either DNA or RNA. Oligoprobes may consist of 17–75 nucleotides, and ribroprobes may be composed of hundreds or thousands of nucleotides. The probe may be labeled with biotin, digoxigenin, fluorescence or radioisotopes. There are DNA-DNA, DNA-RNA, RNA-RNA and RNA-DNA *in situ* hybridizations depicting the various types of probes and targets concerned. An inherent limitation to this technique is that it requires at least 20 copies of identical DNA or RNA sequence in a single cell for detection (44). For this reason, most *in situ* hybridizations have been employed to study mRNA, which is usually present in much higher copy numbers per cell than DNA. In many instances, this leaves a lot to be desired as low copy numbers of DNA or RNA are important for many normal and pathologic conditions. Latent viral infections, for example, often have only a few copies of viral genomes per cell. Gene mutations, chromosomal translocation, gene therapy and early pathological changes in DNA and RNA may all involve fewer copies of nucleotide sequences than those detectable by conventional *in situ* hybridization. Despite their importance, the cellular distribution of these small quantities of nucleotide sequences remained invisible until *in situ* PCR is deployed.

PRINCIPLES AND STEP-BY-STEP CONSIDERATION OF *IN SITU* PCR

Ever since the development of liquid-phase PCR, there have been many attempts to perform *in situ* PCR. Haase and associates were the first to report a successful experiment in 1990 (26). They experimented with fixed intact cells in suspension and successfully performed PCR using the cell nuclear membrane as a sack to retain the amplificants. Enzyme digestion was used to permeate the membranes allowing primers, polymerases and free nucleotides to enter into the nucleus. Then, they managed to enlarge the size of the amplified signals by using multiple primers with complementary tails. The amplified and self-nested products were too large to leak out of the nucleus. The amplification was performed in test tubes and the cells were then spun down on glass slide for detection. Since then, many modifications and advances have been made. Now, more than a dozen laboratories have reported successful applications of *in situ* PCR. In the first issue of *Cell Vision*, many articles relating to this technique were published—most of them using protocols with varying modifications (4,8,11,29,38,61,68,70,71,74–80). Overall, this technique is still in its developmental stages, undergoing multiple phases of advancement. For the most part, many groups concur with the basic principles and the key steps in the procedures. The basic principle and practical considerations for *in situ* PCR are discussed below. The actual protocols can be found in later chapters of this book and in a number of articles of *Cell Vision*.

There are already a number of variations of *in situ* PCR and, undoubtedly, there will be many more. However, the basic steps can be illustrated in Figure 1. A successful *in situ* PCR requires that each step be performed properly and that specificity be verified with appropriate controls. In order to do so, one needs to understand the theoretical background behind each step so that optimal procedures can be established and adapted to suit specific needs.

Tissue Fixation

In situ PCR can be performed on intact cells in suspension, cells in smear, cells in cytospin, metaphase chromosomes, frozen sections and paraffin sections (1,10,15,22,29,34,39,74–80). Ideal fixations for *in situ* PCR should preserve both DNA or RNA and tissue morphology. If immunocytochemistry is to be performed in conjunction with *in situ* PCR on the tissue sample, antigenicity must also be preserved. Fortunately, formalin and paraformaldehyde, the most commonly used fixatives for histopathology and immunocytochemistry, meet these criteria. Fresh tissue is preferred but formalin-fixed archival tissue samples can also be used. Whole cells with intact membranes are ideal as they should have less damaged nucleotide sequences, and their nuclear or cell membrane serves as a natural boundary to retain the amplified products (26,32,44,57). *In situ* PCR on tissue sections is less efficient, but it acquires a higher significance as most pathological tissue samples are fixed in formalin and kept in paraffin blocks. It demands more technical refinement in conducting the procedure. For either paraffin or frozen tissue sections, thicker sections may yield better results. As the target DNA or RNA are molecules of long strands and somewhat randomly distributed in the nuclei or the cytoplasm, a thicker section will contain more target DNA or RNA sequences in their entirety. The depth of the section with more membrane structure and the cross-linked proteins might also serve to trap the amplified products from diffusion. Depending on the availability of the target DNA or RNA, thinner sections (< 7 µm) may also be used, which generally give lower background and better morphology. Immediate fixation in 10% buffered formalin, pH 7.0, for 4 to 6 h at 4°C is an acceptable fixation method. Formalin causes cross-linking between proteins and nucleotides, which may serve to retain the amplified sequences and keep them from being washed away. Unbuffered formalin is not recommended as it has been shown to reduce reaction intensity of conventional *in situ* hybridization (43,54). Alcohol fixation or unfixed tissue samples have also been experimented, but with limited success (10,58). Fixatives with picric acid such as Bouin's fixative or heavy metals such as mercury in Zinker's solution are not recommended as they degrade DNA or RNA after a few hours of fixation (23,24). To test the suitability of a fixative for *in situ* PCR, it is advisable to dissect a portion of the fixed tissue sample and extract DNA or RNA from it. The availability of total DNA or RNA and of the nucleic acid sequence of interest in the fixed tissue sample can be evaluated using conventional PCR, electrophoresis and blotting assays.

Tissue Pretreatment

For *in situ* PCR, it is almost always necessary to digest the tissue sample with a protease, such as proteinase K, trypsin or pepsinogen. It permeates the tissue sample, allowing reagent penetration and unveiling the target sequence for amplification. The concentration of the enzyme and the duration and temperature of the treatment are important. Excessive digestion distorts the tissue morphology and increas diffusion of PCR products through disrupted membrances of intact cells (34,39,44). If under-digested, the tissue will have poor permeability and extensive cross-linking between proteins and nucleotides, which interfere with PCR (44). Both may lead to false negative results. The strength of digestion should also correlate with the length of fixation. The more extensively the tissue sample is fixed, the heavier the digestion should be. After digestion, the enzyme should be completely inactivated by heating or removed by washing. Minute remnants of the enzyme may destroy the DNA polymerase, which is essential for PCR amplification.

Primer Design

Primers are usually 18–28 nucleotides long and encase a fragment of around 100–1000 bp. For *in situ* PCR, shorter template sequences are preferred. This is especially true for archival tissue samples in paraffin sections as considerable degradation of DNA and RNA have taken place. DNA fragments extracted from paraffin blocks are rarely longer than 400 bp, and RNAs are not more than 200 bases (63). Long sequence amplifications are more prone to nonspecific reactions caused by mispriming on the original DNA (13). Once a wrong sequence has been copied, it will be continuously amplified by the remaining cycles of PCR. It is important to design a pair of primers that has little or no homology to any other sequences in the tissue. They should also have no homology to each other or within themselves. Computer programs are available to verify the degree of homology for each designed fragment. Nonspecific or false positivity generated by PCR is, to a large extent, attributable to mispriming of the primers (12). Generally, one pair of primers will be sufficient. To increase specificity of the amplification, multiple pairs may be used (10,14–16,26,33,39,49,64). A second pair of primers embedded within the sequence amplified by the first pair can be employed. This reaction is called "nested PCR." It allows further amplification of the signal already amplified by the first pair to increase specificity and sensitivity. Multiple pairs of primers, distinct from each other and against different portions of the same gene sequences, may also be used to increase the specificity of the detection.

PCR

PCR performed in tissue sections or intact cells follows the same principles as conventional PCR. One important difference between *in situ* PCR and liquid-phase PCR is that DNA or RNA strands in fixed cells or tissue sections are immobilized. Not all the target sequences are available for complementary bonding and, more likely than not, the entirety of the target sequence has been damaged or truncated during tissue processing and sectioning. Twenty-five to forty cycles are recommended, although the *in situ* amplification reaction may have reached its plateau in less than 20 cycles (1,34,39,52,58,62). *In situ* PCR can be performed on a conventional PCR thermocycler by using aluminum foil paper to make a flat plate out of the platform. The temperature is effectively transmitted through the glass slide to the tissue sample and reaction solution. Mineral oil may be used to facilitate heat conduction. However, conventional PCR machines may only hold about four slides for each run, which is not nearly enough for the many controls necessary for each experiment. Thermocyclers specifically designed for *in situ* PCR have been available. They can hold from 12 to 20 slides per round and are much more efficient for carrying out experiments. Typically, the duration of each temperature step for *in situ* PCR is slightly longer than that of conventional PCR to overcome the disadvantages inherent in *in situ* PCR and to ensure that sufficient amplification takes place (44). Denaturing at 94°C for 1 min, annealing at 55°C for 30 s and extension at 75°C for 1 min are generally adequate. The *Taq* DNA Polymerase has a half life of about 20 min at 96°C (59). Prolonged total time of heating will diminish its activity. The concentrations of primers, DNA polymerase and magnesium may also be higher than those used in conventional PCR (32,39,49). For tissue sections, cytospins and smears, the amplification is performed on glass slides. Generally, it takes place under a coverslip that is sealed with nail polish or a Pap pen (Newcomer Supply, Oak Park, IL, USA) (69,78).

It is commonly believed that *in situ* PCR is not as efficient as conventional PCR. It was estimated that liquid-phase PCR can amplify a single copy of DNA to billions of identical copies while *in situ* PCR amplifies to much fewer copies (44). This estimation was performed by measuring the end products of the liquid supernatant or homogenate of the tissue samples at the end of *in situ* PCR thermocycling (67). The concentrations of amplified products, however, was calculated against the total tissue volume or liquid volume which, by any measure, is not a fair denominator. The true concentrations of the end products should be calculated against the value of the individual cells that harbor the products. For conventional PCR, as the cycle number increases, amplification efficiency is affected by substrate saturation and exhaustion of reactive solution. At a certain point of amplification, the exponential growth plateaus (13). This effect is influenced by template input and initial amplification efficiency. Over how many cycles this phenomenon occurs for any unknown concentration of template is not completely predictable (36). For *in situ* PCR, it is presumed that this effect comes much earlier and is mainly attributed to the local concentration of the amplificants. This local concentration for *in situ* PCR is likely to be as high as the total concentration of liquid-phase PCR. Detecting sensitivity generally refers to the minimum copy numbers of a target detectable by the method. If both *in situ* PCR and liquid-phase PCR detect down to a single copy of DNA or RNA, they are of equal detecting sensitivity.

A "hot start" technique that increases the specificity of *in situ* PCR has been developed (44,47,49). Initial annealing between the primers and the target sequence determines the amplification specificity. "Hot start" entails the initiation of the primer-target annealing step at a higher temperature, which significantly reduces the possibility of mispriming, and thereby improves the specificity of subsequent PCR (17). There are a number of variations in performing the "hot start." For conventional PCR, a special wax called AmpliWax (Perkin-Elmer, Norwalk, CT, USA) can be applied to separate the reaction mixture and the polymerase. It serves as a barrier between the reagents but will melt at temperatures above 80°C. The DNA polymerase will then be mixed with the reaction mixture as the wax melts, ensuring that PCR starts at a high temperature, a condition strongly unfavorable to nonspecific annealing. For *in situ* PCR, "hot start" requires adding the DNA polymerase after the slides have been heated to about 94°C for a few minutes. Now there is specific monoclonal antibody available that specifically reacts to *Taq* DNA Polymerase and effectively blocks the enzymatic activity up to 70°C. This inhibition is completely reversed upon the first denaturation step in thermal cycling when the antibody is inactivated by heating (Clontech Laboratories, Palo Alto, CA, USA). It has been confirmed that "hot start" is crucial for a highly specific and efficient *in situ* PCR. Once the PCR starts with specific initial annealing and amplification, the rest of the duplication cycles will usually yield products of high fidelity.

A very important step for a successful *in situ* PCR is product retention. For intact cells, the amplified products may be withheld by cellular and nuclear membranes (26,57). For tissue sections, however, it is not entirely clear why the large number of amplified sequences stay *in situ*, nor is it known how to make them stay at the site of the original template. It is reasonable to believe that most of the amplified sequences have diffused away following established rules of physics. The well-defined distribution of the reaction, as visualized by the detection system, would strongly indicate that certain mechanisms exist which trap the amplified sequences in place. This may, in fact, be the case. During PCR, while the sequences encased by the two primers increase exponentially with each cycle, there are also "long products" that increase in number arithmetically. These long products are generated by duplication of the original template, and their chains extend beyond the boundary of the primer on the opposite strand. Two more such byproducts are produced by each PCR cycle. After 30 cycles, at least 60 such long sequences are generated for each copy of target DNA. They should be able to bind to the intrinsic templates that are partially embedded within the tissue section and also to the short sequences and to each other as they share complementary sequences. The long products could act as anchors, effectively retaining some of the amplified products. Judging from the appearance of the final reaction in the *in situ* PCR preparation, it is possible that those sequences retained by the long products are the major sources of signals that are not diffused or washed away during the procedure. Sixty or more identical target copies retained by the long products should be easily detectable by subsequent *in situ* hybridization and give a well-defined signal made visible by the detecting system. The long products, that are mostly neglected and have little importance in conventional PCR, may be the most important products of *in situ* PCR. This hypothesis is illustrated in Figure 1 (I).

Other measures have been taken to increase the chance of product retention. Thicker sections with deeper spatial structure may serve to trap the amplified sequences in place. Sealed coverslips or plastic bags have been employed to prevent evaporation and also to retain the products *in situ*. Post-PCR drying at about 60°C and fixation in 4% paraformaldehyde are also necessary to effectively immobilize the amplified products (10,15,26,78). It is possible that these measures do not work by themselves but only facilitate the bonding of the long products to the amplified signals described above.

Following PCR, the samples should be washed gently with solutions of appropriate stringency. Many investigators have employed different strategies in this washing step. Excessive washing may remove the amplified products from the initial site. However, without washing, high background or diffused signals may be seen. The washing stringency can be designed according to the T_m formula described previously. It is sometimes a compromise between the amounts of desired specific signal retention and that of background nonspecific staining.

In Situ Hybridization Detection

The amplified signals are detected by *in situ* hybridization. The initial step normally starts with pre-hybridization, which entails incubating the sections with all the chemical ingredients in the hybridization solution except for the specific probe. Those ingredients will saturate the potential nonspecific bonding sites on the tissue sample to increase the specificity of the detection (66). A pre-hybridization step can be performed before PCR (78,80). This modification is commendable as it will reduce the steps and number of washings after PCR and serve to better preserve the amplificants. An enzyme digestion routinely used in conventional *in situ* hybridization is not necessary for PCR product detection. The *in situ* hybridization can be carried out in a number of ways (27). The most straightforward is to employ an oligoprobe that is labeled with biotin or digoxigenin and hybridized under appropriate conditions with formamide at a temperature of 37°–54°C from 3 h to overnight. Generally, a 3-h incubation at 42°C is sufficient. Longer probes, such as riboprobes, with sequences extending beyond the fragment that is amplified, may be employed. Longer probes increase the detecting specificity but decrease the sensitivity (66). Multiple probes can also be used to ensure that different fragments of the amplified signal correlate well with the sequence of interest. The design of a specific probe is very important as PCR may amplify more than one sequence and a specific *in situ* hybridization will pick up only the correct one (13). Different labeling methods are more a matter of preference than a limiting factor. Generally, radioisotope labeling gives the highest detecting sensitivity but it is compromised by time, hazardous chemicals and reduced resolution (27,28,66). Fluorescent labeling is also very sensitive. However, its signal tends to fade and the background structure is not clearly visible. Protein labeling with enzyme detection is at present the method of choice (27,28). Biotin labeling can be easily detected by the streptavidin-biotin complex that can be coupled to a number of enzymes, such as alkaline phosphatase or horseradish peroxidase. Digoxigenin labeling can be recognized by a specific antibody that will have added specificity.

The detection procedure of hybridization signal is very similar to immunocytochemistry. It may have many variations ranging from direct visualization to immunogold-silver autometallography and display a variety of colors, intensity and textures (27). The sections of tissue can then be counterstained and examined.

Controls

The entire *in situ* PCR procedure is fairly long and technically demanding. Each inappropriate step may create false positive or negative and should be properly controlled. In theory, there are more than 20 controls that should be run with each experiment so that the negative or positive results can be ascertained (13,32,44). If any of the controls give unexpected results, one should be able to explain and define the causes of the problem and interpret the results accordingly. However, in reality, the following controls should be routinely performed as the first line checklist to ensure the specificity of the reactions. Usually, unrelated primers or omission of the primers should be used as controls. This should give a negative or much weakened signal. The use of unrelated probes or omission of the probes should be the best control for the *in situ* hybridization step. It is important to have good positive and negative samples to run in each experiment to make sure that all the ingredients and steps are properly applied. A negative control, for example, could be a tissue pretreated with DNase or RNase to destroy the targets before being subjected to the *in situ* PCR experiment. When intact cells in suspension are used, artificial mixtures at different ratios with another cell type known not to contain the target sequence provide good controls (1,32,44). The cell preparation may be smeared, cytospun or grown on slides if they attach. They also may be collected into pellets by centrifugation, embedded or frozen and cut, and then serve as controls for tissue sections. One good working model is SiHa cells, a cell line derived from human cervical cancer cells. It is known to contain a single copy of human papilloma virus (HPV) type 16 per cell (44,52,71). Other cell lines such as HeLa cells, known to contain about 25 copies of HPV type 18 per cell, and Caski cells, known to contain about 600 copies of HPV 16 per cell, are also good controls for technical setup. DNA or RNA extracted from the same tissue sample should be assayed with conventional PCR to verify the presence or absence of the sequences of interest. If conventional PCR results do not correlate with those of *in situ* PCR, a check on the fixation or enzyme digestion may be in order. The washing stringency may be adjusted to facilitate a better retention of the amplified signal or a reduction in background staining. The amplification solution should also be examined after the PCR by recovering a small amount of the reagents from the top of slides and running it through liquid-phase PCR or directly through Southern or Northern blotting to check for the presence of the amplified sequence of interest (67). The results normally provide information that will clarify at which steps the procedure went wrong and thereby help to troubleshoot the problem. If there is too much background or nonspecific reaction, adjustments may be made to reduce the PCR cycles, redesign the primers, modify the concentrations, change the annealing temperature, perform the "hot start" carefully, block nonspecific reactions in the detecting steps, increase the washing stringency, etc. It also should be kept in mind that as the detecting sensitivity of *in situ* PCR surpasses other detecting systems, unexpected positivity may reflect the true presence of the sequence in a particular tissue structure. This is true for HIV and a number of other viruses that were not known to be present in a variety of cell types. Their latent phase of infection is now gradually being unveiled by *in situ* PCR (1,3,5,15,16,31,53,57,74–80).

Most of the problems and pitfalls encountered in conventional PCR and *in situ* hybridization also need to be dealt with in *in situ* PCR. The extensively published protocols and modifications on optimizing and troubleshooting conventional PCR are often applicable to *in situ* PCR. PCR product contamination, a nightmare for conventional PCR, is also a concern for *in situ* PCR. A dedicated working area that is nucleic acid-, RNase- and DNase-free is highly recommended. Nevertheless, most of the false positive or negative results for *in situ* PCR are not caused by contamination but by mispriming, inadequate primer design, inappropriate fixation or pretreatment and product diffusion. False positivity in intact cell preparation may be caused by leakage of the amplified products that may serve as templates for *in situ* PCR amplification at another site. More attention should be paid to these aspects to establish and optimize the protocol in each particular case.

Major Variations of *In Situ* PCR

A number of important modifications and variations have been introduced for *in situ* PCR during its short history of development (Figure 2). Indirect *in situ* PCR, as discussed above, entails a PCR followed by *in situ* hybridization that detects amplified signals with specific probes. It has been the most widely used procedure up to now (1,3,5,10,26,34,39,49,52,64,68–70,72–80).

The second important variation is direct *in situ* PCR, which uses primers or free nucleotides labeled with proteins or other labeling reagents (15,16,29,34,39,52,57,62). The labeling is built into the amplified products and a direct visualization is possible without any subsequent *in situ* hybridization. This procedure is more straight-

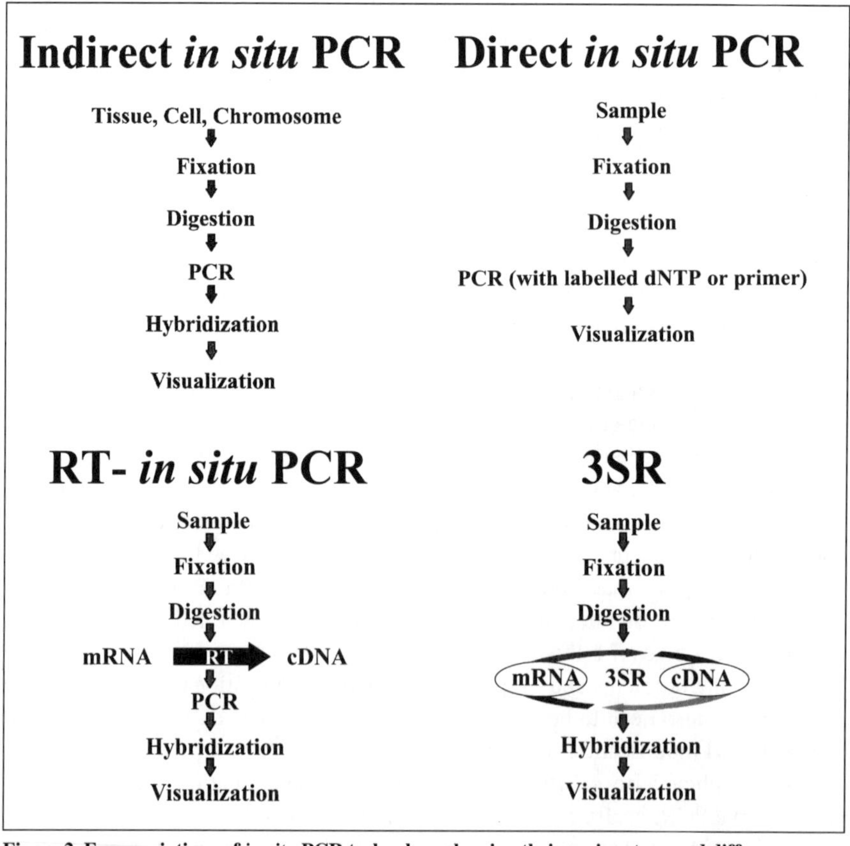

Figure 2. Four variations of *in situ* PCR technology showing their major steps and differences.

forward and easier to perform. However, there has been concern about the specificity of the reaction as many misprimings or nonspecific annealing may occur in the procedure (72). Nonspecific labeling may be caused by the repair process to damaged DNA. This is more prevalent in tissue sections than in fresh DNA extracts (63). Direct *in situ* PCR also lacks the subsequent *in situ* hybridization, which serves as an additional check on specificity by selecting only the particular amplificants of interest. Labeled primers also tend to render the PCR less efficient (13). Direct *in situ* PCR may be applied to intact cell preparations or tissue sections. With highly specific primers and target sequences and well-controlled reaction conditions, direct *in situ* PCR can yield meaningful results. It remains an important variant.

RNA can also be detected by *in situ* PCR with an added reverse transcription (RT) step. This process is called *in situ* RT-PCR (15,29,35,44,50,56,57,61,74,75,77). *In situ* RT-PCR requires an initial treatment of the tissue sample with DNase to destroy the intrinsic DNA in order to ascertain that all amplified signals derive from reverse transcribed RNA. This DNase treatment is not necessary if the two primers are known to be far apart enough on the genomic DNA so that no residual DNA will be amplified. The first strand of cDNA is synthesized from the RNA template by a reverse transcriptase, random hexamers and free nucleotides at about 42°C for 30 to 60 min. Once the cDNA is made, the reverse transcriptase can be inactivated by heating at above 90°C. The newly formed cDNA is then used as the template for amplification. A new enzyme, rTth (Perkin-Elmer, Norwalk, CT, USA), is available that can perform both the reverse transcription and the subsequent PCR with the same enzyme and shortens the procedure. Variations in primer design, multiple primers and nested primers are also viable alternatives to improve specificity and efficiency of *in situ* RT-PCR.

A newly developed method for RNA detection, named self-sustained sequence replication reaction (3SR), was reported by Ingeborg Zehbe and colleagues (70,71). They adapted an enzyme-driven *in vitro* mRNA amplification method (19–21,25) and successfully applied it to intact cells in cytospin preparations. This method circumvents the PCR protocol. It utilizes three enzymes, i.e., avian myeloblastosis virus (AMV) reverse transcriptase, T7 RNA polymerase and *Escherichia coli* RNase H, to induce reiterated cycling of RT and transcription reactions to replicate RNA target via RNA/DNA hybrids and double-stranded cDNA intermediates. The cDNA copy of the original RNA contains a phage (T7) RNA polymerase promoter that is used as a template for an approximately 10^7-fold amplification within 1–2 h. All the reactions are carried out at 42°C and no thermocycler is necessary. It provides an alternative to *in situ* RT-PCR for low copy number RNA detection in intact cells or tissue sections. It is particularly useful in combining with immunostaining as the antigenicities are better preserved without the high temperature exposure in heated PCR. Increasingly, *in situ* PCR and its variants are used jointly with other staining methods, such as immunohistochemistry, to facilitate a better analysis of tissue components under investigation.

The possibility of performing *in situ* PCR on chromosomal preparations has been demonstrated by Gosden and Hanratty, although they used only a single primer in the experiment (22). *In situ* PCR has also been attempted at the electron microscopic level; however, the results were very preliminary (73). Ultrastructural *in situ* PCR remains a possibility awaiting further experimentation.

Applications of *In Situ* PCR

The application of *in situ* PCR has been mainly confined to two areas: foreign gene detection and identification of gene alterations, although many other areas remain to be fully explored (Figure 3). Because of its extremely high sensitivity, *in situ* PCR for DNA is limited to the detection of gene sequences that are not normally present in the tissue samples of interest. Foreign gene detection can be further divided into two categories, i.e., detection of infectious agents such as bacteria, fungi and, particularly, viruses, and detection of artificially introduced genes such as in gene therapy, transgenic animals and graft transplants. The ability to detect low copy numbers of viral genes has led to a number of significant discoveries of latent phases of viral diseases, particularly AIDS, hepatitis and herpes. With this technique, the knowledge gap in the pathogenesis of early and latent viral infection is being filled.

The employment of *in situ* PCR has led to major advances in AIDS research. Bagasra and associates analyzed peripheral monocytes of eleven HIV seropositive patients by means of *in situ* PCR (2). Eight were found to contain HIV sequences that were not detectable by conventional *in situ* hybridization. The peripheral blood mononuclear cells in 56 HIV seropositive and 11 seronegative patients were further analyzed with *in situ* PCR (1). It was found that 0.1%–13.5% of the cells harbored

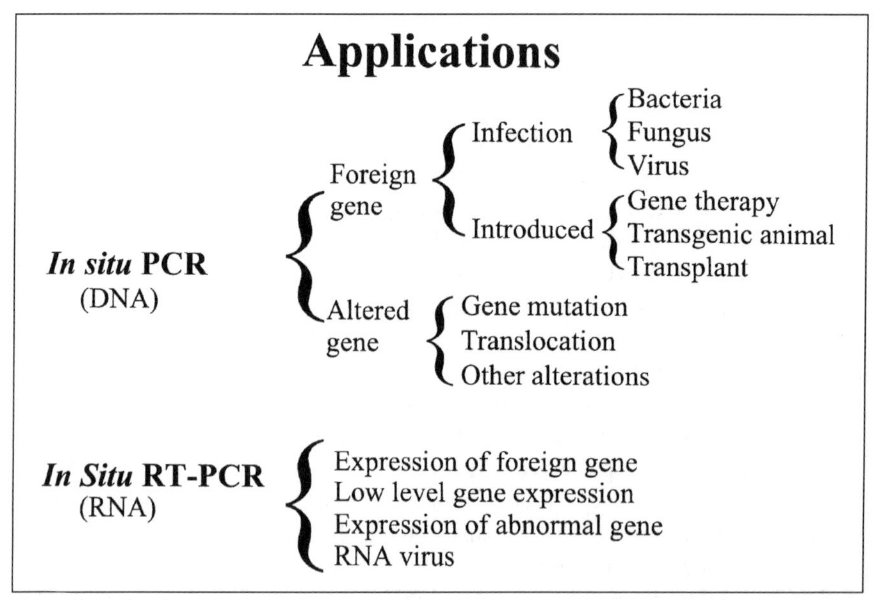

Figure 3. Major applications of *in situ* PCR by category. Most of the potential applications have just begun to emerge or have not been fully explored.

HIV genome sequences and the viral load correlated with the stages of the disease. No positivity was detected in the HIV seronegative subjects. The CD4 subpopulation of the T lymphocytes were then examined in 42 HIV-1 infected patients (3). From 0.2%–69% of the cells were found to harbor HIV-1 sequences, and the percentages increased significantly with advancing stages of disease. Similar observations were made, using direct *in situ* PCR, by Nuovo et al. who detected HIV-1 DNA in peripheral leukocytes of all 10 AIDS patients studied (44). The positive cells accounted for 1.8%–6.2% of the peripheral lymphocytes. Patterson and associates performed *in situ* RT-PCR in blood cells of HIV-1 infected patients in suspension (57). The amplified signals were hybridized to fluorescent-labeled probes, and the cells were then sorted with flow cytometry into HIV-1 positive and negative cells. Analysis of proviral DNA and mRNA showed that the HIV-1 genome persists in a large reservoir of latently infected cells. The number of peripheral lymphocytes that contain HIV sequences were manyfold higher, as reported in these studies employing *in situ* PCR, than what had been reported previously. *In situ* PCR of body fluid also offers the possibility of identifying HIV infection before seroconversion or in infants of HIV-infected mothers before the infants are 18 months of age. At that age, the inherited HIV-antibodies clear away from the body. Embretson et al. used *in situ* PCR of HIV sequences to examine CD4 cells in the lymph nodes of AIDS patients (16). A very large number of latently infected CD4 lymphocytes and macrophages throughout the lymphoid system, from early to late stages of infection, were found to contain the virus. Extracellular association of HIV with follicular dendritic cells was confirmed. These cells may transmit infection to other cells as they migrate through lymphoid follicles. The same authors also used *in situ* PCR and *in situ* RT-PCR comparatively in routinely fixed and paraffin-embedded tissues from AIDS patients (15). In one tumor biopsy, many of the lymphocytes and macrophages, some infiltrating the tumor, were found to contain HIV DNA. Fewer than one in a thousand cells contained HIV RNA, indicating that most of these cells were latently infected. Nuovo et al. examined HIV-1 DNA and RNA in cervical tissues of women with AIDS by employing *in situ* PCR and *in situ* RT-PCR (46). Amplified HIV-1 DNA and cDNA were detected in each of the 21 cervical biopsies. They were most abundant in the endocervical aspect of the transformation zone at the interface of the glandular epithelium and the submucosa and in the deep submucosa around microvessels. Many positive cells were identified as activated macrophages. These HIV-1 harboring cells in the cervix may represent primary HIV infection sites for these women. Using *in situ* PCR, Zevallos et al. investigated the placentas of 48 HIV-1-infected mothers (79). HIV-1 sequences were detected in 27 (56%) of the cases but in none of the 10 HIV seronegative cases. The results obtained with *in situ* PCR and conventional PCR correlated well. The HIV-1 sequences were identified in a number of cell types in the placenta that were not previously known to be infected by HIV, including syncytotrophoblasts and villous-stromal cells (Hofbauer cells). The findings provided clues to the mechanism of mother-to-fetus HIV transmission. Employing the same technology, the same authors also identified HIV-1 sequence harboring cells in the lungs (75,76) and lymphoid tissue (78) of AIDS patients. The above findings could not have been made without using *in situ* PCR.

Together, these findings unveil a new picture of HIV infection. The possible pathogenesis of a number of complications of AIDS is beginning to emerge.

A number of other viral infections, including HPV, HBV, HCV, CMV, HTLV-1, Visna-virus and MMTV provirus, have been studied with *in situ* PCR or *in situ* RT-PCR (8,29,45,48,52,55). Nouvo et al. examined HPV, measles and hepatitis C in archival tissue samples and cell lines. HPV types 6 and 11 in penile lesions detected by *in situ* PCR were found in cells that lacked halos and atypia, which were the conventional diagnostic features for condyloma/low-grade intraepithelial lesions (45). They also used *in situ* PCR to examine HPV 16 in formalin-fixed and paraffin-embedded cervical squamous intraepithelial lesions (52). The viral sequence was detected in the basal cells and parabasal cells at the site of the lesion that have only rarely been detected by conventional *in situ* hybridization. Isaacson et al. examined polio, measles, influenza and HTLV-1 in archival brain tissue samples with *in situ* RT-PCR and detected their RNA sequences in individual neurons, glia and/or vascular endothelial cells (29). Some of the paraffin tissue blocks were over 25 years old.

In situ PCR has been employed to study gene rearrangement, mutations and chromosomal translocations (14,22,34,37,39,56,58,65). In this regard, much still remains to be explored. For example, gene alterations in oncogenesis could be a good indicator for tumor type and prognosis. *In situ* PCR will also find wide application in the study of genetic diseases. The specific DNA sequence alteration, both acquired and inherited, and low-level gene expression underlying specific genetic disorders can be identified in individual cell types with this technique. Cells with defective genes and their expression can be pinpointed at different developmental stages, providing a new foundation for the understanding of many genetic disorders. They can also be used in tracing the genomic damages caused by different diseases, therapies or pollution. Embleton et al. used *in situ* RT-PCR of immunoglobulin variable region genes in two mouse hybridoma cell lines and the bcr-abl fusion genes in human K562 myeloid leukaemia cells (14). The amplified Ig VH and VL DNA could be assembled within the same cells using suitable PCR primers. The experiment indicates the possibility of constructing human antibodies *in vitro*.

In situ PCR will also be a powerful tool to assist gene therapy in tracing the whereabouts of "foreign" genes and pinpointing the exact locations of the introduced gene sequences after gene delivery (68). This can be studied at both the cellular level and the chromosomal level. This method will compensate well for the currently employed marker genes, such as *lacZ* coding for beta-galactosidase or Agfa coding for alkaline phosphatase, which have their inherent shortcomings in gene therapy. Yin and associates infused preproenkephalin promoter genes with a bacterial *lacZ* marker in the rat brain (68). The location of the introduced gene was detected at three levels, i.e., by histochemical stain for the protein beta-gal, by *in situ* hybridization for mRNA and by *in situ* PCR for the *lacZ* gene itself. The approach is meaningful as not all the genes internalized by the target cells are expressed, nor are all the gene transcripts translated. The information concerning the exact location of the introduced gene is extremely valuable for improving the knowledge of gene therapy, an area that is rapidly gaining investigative significance and will in the near future have increasingly profound clinical and therapeutic implications.

While *in situ* PCR of DNA is mainly restricted to detection of foreign or altered gene sequences that do not normally exist in the cells, *in situ* RT-PCR can be applied to any gene expression that is beyond the detection limit of conventional *in situ* hybridization (15,29,44,51,56,57,61,74,75,77). Patel et al. detected epidermal growth factor receptor mRNA in tissue sections from biopsy specimens using *in situ* RT-PCR with digoxigenin-labeled dUTP (56). Zevallos et al. detected mRNA of retinoblastoma in cultured leukocytes and fetal lung (74,75,77). Smith-Norowitz et al. detected beta-actin mRNA and IL-6 mRNA in unstimulated spleen cells of BALB/c mice (61). Currently, *in situ* RT-PCR and *in situ* 3SR are the only morphological means of visualizing low copy numbers of mRNA in particular cell types. Although the genomic DNA with a hundred thousand or more coding sequences is universally present in almost every cell, the mRNA is not. Only those genes that are transcriptionally active produce corresponding mRNA. Many quiescent genes are known to produce small amounts of mRNA. Their significance has not been elucidated. *In situ* RT-PCR will be able to detect their expression by reverse transcribing the mRNA into cDNA and then amplifying the cDNA *in situ*. Up to now, there have been only limited applications of *in situ* RT-PCR. The possibilities of applying it to study low copy number gene expression await exploration and present new horizons.

Quantification of *in situ* PCR remains a possibility, but it is still at a premature stage. Most morphologic quantification is based on the counting of positive cells or the measurement of staining intensity, which is not nearly as straightforward as it would appear. A considerable amount of borderline positivities/negativities often leaves room for subjectivity on the part of the investigator. Computerized image analysis may reduce but not eliminate this problem. Possible pitfalls of *in situ* PCR quantification include variations in every step of the protocol, especially plateau effect and product diffusion. Owing to its amplification power plus product diffusion and the plateau effect, a cell containing one copy of a gene and another containing 10 copies of the same gene may present similar reaction intensities following 10 cycles of PCR. At present, cells in suspension are the best candidates for attempts at quantification (1,3,5,57). It is possible to incorporate radioisotope-labeled free nucleotides in PCR and use autoradiography to assess the amounts of incorporation at the target sites to give a quantitative representation of the local reaction. However, given the problems associated with direct *in situ* PCR, the results need to be interpreted with caution. For tissue sections, quantification of *in situ* PCR results is not recommended at this time (39,40). Before more precise knowledge has been obtained, quantification of *in situ* PCR remains, at best, comparative and semiquantitative.

Like any new, technically challenging technology, *in situ* PCR is going through a period of trial and error. However, judging from the results obtained so far and in theory, this technique will be one of the most promising and powerful morphologic technologies that we have witnessed. It is likely to provide insight into a number of normal and pathological mechanisms. For those who need to trace minute quantities of foreign genes or altered genes, or to study low copy numbers of RNAs in the background of cell morphology, *in situ* PCR and its variants provide the only tool available to answer the impending questions.

REFERENCES

1. **Bagasra, O., S.P. Hauptman, H.W. Lischner, M. Sachs and R.J. Pomerantz.** 1992. Detection of human immunodeficiency virus type 1 provirus in mononuclear cells by *in situ* polymerase chain reaction. N. Engl. J. Med. *326*:1385-1391.
2. **Bagasra, O. and R.J. Pomerantz.** 1993. Human immunodeficiency virus type I provirus is demonstrated in peripheral blood monocytes *in vivo*: A study utilizing an *in situ* polymerase chain reaction. AIDS Res. Hum. Retroviruses *9*:69-76.
3. **Bagasra, O., T. Seshamma, J.W. Oakes and R.J. Pomerantz.** 1993. High percentages of CD4-positive lymphocytes harbor the HIV-1 provirus in the blood of certain infected individuals. AIDS *7*:1419-1425.
4. **Bagasra, O., T. Seshamma and R.J. Pomerantz.** 1994. *In situ* polymerase chain reaction: Applications in the pathogenesis of diseases. Cell Vision *1*:48-51.
5. **Bagasra, O., T. Seshamma and R.J. Pomerantz.** 1993. Polymerase chain reaction *in situ*: Intracellular amplification and detection of HIV-1 proviral DNA and other specific genes. J. Immunol. Methods *158:*131-145.
6. **Bolton, E.T. and B.J. McCarthy.** 1962. A general method for the isolation of RNA complementary to DNA. Proc. Natl. Acad. Sci. USA *48*:1390.
7. **Bonner, T.I., D.T. Brenner, B.R. Neufeld and R.J. Britten.** 1973. Reduction in the rate of DNA reassociation by sequence divergence. J. Mol. Biol. *81*:123-135.
8. **Cartun, R.W., J.F. Siles, L.M. Li, M.M. Berman and G.J. Nuovo.** 1994. Detection of hepatitis C virus infection in hepatectomy specimens using immunohistochemistry with reverse transcriptase (RT) *in situ* polymerase chain reaction (PCR) confirmation. Cell Vision *1*:84.
9. **Casey, J. and N. Davidson.** 1977. Rate of formation and thermal stabilities of RNA:DNA and DNA:DNA duplexes at high concentrations of formamide. Nucleic Acids Res. *4*:1539-1552.
10. **Chieu, K.-P., S.H. Cohen, D.W. Morris and G.W. Jordan.** 1992. Intracellular amplification of proviral DNA in tissue sections using the polymerase chain reaction. J. Histochem. Cytochem. *40*:333-341.
11. **Cirocco, R., M. Careno, C. Gomez, K. Zucker, V. Esquenazi and J. Miller.** 1994. Chimerism demonstrated on a cellular level by *in situ* PCR. Cell Vision *1*:84
12. **De Bault, L.E. and B.-L. Wang.** 1994. Localization of mRNA by *in situ* transcription and immunogold-silver staining. Cell Vision *1*:67-70.
13. **Eeles, R.A. and A.C. Stamps.** 1993. Polymerase Chain Reaction (PCR). The Technique and Its Applications. R.G. Landes Company, Austin, TX.
14. **Embleton, M.J., G. Gorochov, P.T. Jones and G. Winter.** 1992. In-cell PCR from mRNA: Amplifying and linking the rearranged immunoglobulin heavy and light chain V-genes within single cells. Nucleic Acids Res. *20*:3831-3837.
15. **Embretson, J., M. Zupanic, T. Beneke, M. Till, S. Wolinsky, J.L. Ribas, A. Burke and A.T. Haase.** 1993. Analysis of human immunodeficiency virus-infected tissues by amplification and *in situ* hybridization reveals latent and permissive infections at single cell resolution. Proc. Natl. Acad. Sci. USA *90*:357-361.
16. **Embretson, J., M. Zupancic, J.L. Ribas, A. Burke, P. Racz, K. Tenner-Racz and A.T. Haase.** 1993. Massive covert infection of helper T lymphocytes and macrophages by HIV during the incubation period of AIDS. Nature *62*:359-362.
17. **Erlich, H.A., D. Gelfand and J.J. Sninsky.** 1991. Recent advances in the polymerase chain reaction. Science *252*:1643-1650.
18. **Gelfand, D.** 1989. *Taq* DNA polymerase, p. 17-22. *In* H.A. Erlich (Ed.), PCR Technology. Stockton Press, New York.
19. **Gingeras, T.R., P. Prodanovich, T. Latimer, J.C. Guatelli, D.D. Richman and K.J. Baninger.** 1991. Use of self-sustained sequence replication amplification reaction to analyze and detect mutations in zidovudine-resistant human immunodeficiency virus. J. Infect. Dis. *164*:1066-1074.
20. **Gingeras, T.R., D.D. Richman, K.Y. Kwoh and J.C. Guatelli.** 1990. Methodologies for *in vitro* nucleic acid amplification and their applications. Vet. Microbiol. *24*:235-251.

21. **Gingeras, T.R., K.M. Whitfield and D.Y. Kwoh.** 1990. Unique features of the self-sustained sequence replication (3SR) reaction in the *in vitro* amplification of nucleic acids. Ann. Biol. Clin. (Paris) *48*:498-501.
22. **Gosden, J. and D. Hanratty.** 1993. PCR *In situ*: A rapid alternative to the *in situ* hybridization for mapping short, low copy number sequences without isotopes. BioTechniques *15*:78-80.
23. **Greer, C.E., J.K. Lund and M.M. Manos.** 1991. PCR amplification from paraffin-embedded tissues: Recommendations on fixatives for long-term storage and prospective studies. PCR Method Appl. *95*:117-124.
24. **Greer, C.E., S.L. Peterson, N.B. Kiviat and M.M. Manos.** 1991. PCR amplification from paraffin-embedded tissues: Effects of fixative and fixative times. Am. J. Clin. Pathol. *95*:117-124.
25. **Guatelli, J.C., K.M. Whitfield, D.Y. Kwoh, K.J. Barringer, D.D. Richman and T.R. Gingeras.** Isothermal, *in vitro* amplification of nucleic acids by a multienzyme reaction modeled after retroviral replication. Proc. Natl. Acad. Sci. USA *87*:1874-1878.
26. **Haase, A.T., E.F. Retzel and K.A. Staskus.** 1990. Amplification and detection of lentiviral DNA inside cells. Proc. Natl. Acad. Sci. USA *87*:4971-4975.
27. **Hacker, G.W., I. Zehbe, C. Hauser-Kronberger, A.H. Graf and O. Dietze.** Detection of nucleic acids by immunogold-silver staining (IGSS), fluorescent-, peroxidase- and alkaline phosphatase-based methods. Cell Vision *1*:71-73.
28. **Hacker, G.W., I. Zehbe, C. Hauser-Kronberger, J. Gu, A.-H. Graf and O. Dietze.** *In situ* detection of DNA and mRNA sequences by immunogold-silver staining (IGSS). Cell Vision *1*:30-37.
29. **Isaacson, S.H., D.M. Asher, C.J. Gibbs and D.C. Gajdusek.** *In situ* RT-PCR amplification in archival brain tissue. Cell Vision *1*:84.
30. **Kawasaki, et al.** 1989. PCR Technology, Principles and Applications for DNA Amplification. H.A. Ehrlich (Ed.) Stockton Press 90.
31. **Kolata, G.** Tests show infection by AIDS virus affects greater share of cells. 1993. The New York Times; January *5*:C3.
32. **Komminoth, P. and A.A. Long.** 1993. *In situ* polymerase chain reaction. An overview of methods, applications and limitation of a new molecular technique. Virchows Arch. [B] *64*:67-73.
33. **Komminoth, P., A.A. Long, R. Ray and H.J. Wolfe.** 1992b. Comparison of *in situ* polymerase chain reaction (*in situ* PCR), *in situ* hybridization (ISH) and polymerase chain reaction (PCR) for the detection of viral infection in fixed tissue. Pathologica *25*:253.
34. **Komminoth, P., A.A. Long, R. Ray and H.J. Wolfe.** 1992. *In situ* polymerase chain reaction detection of viral DNA, single copy genes and gene rearrangements in cell suspensions and cytospins. Diagn. Mol. Pathol. *1*:85-97.
35. **Komminoth, P., F.B. Merk, I. Leav, H.J. Wolfe and J. Roth.** 1992. Comparison of 35S and digoxigenin-labeled RNA and oligonucleotide probes for *in situ* hybridization. Expression of mRNA of the seminal vesicle secretion protein II and androgen receptor genes in the rat prostate. Histochemistry *93*:217-228.
36. **Larzul, D., F. Guigue, J.J. Sninsky, D.H. Mach, C. Brechot and J.-L. Guesdaon.** 1988. Detection of hepatitis B virus sequences in serum by using *in vitro* enzymatic amplification. J. Virol. Methods *20*:227-237.
37. **Long, A.A.** Detection of gene rearrangements by *in situ* PCR. Sixth annual workshop on recent advances in molecular pathology. Tufts, New England Medical Center, Boston, MA, April 3-6, 1991.
38. **Long, A.A. and P. Komminoth.** 1994. Study of viral DNA using *in situ* PCR. Cell Vision *1*:56-57.
39. **Long, A.A., P. Komminoth, E. Lee and H.J. Wolfe.** 1993. Comparison of indirect and direct *in situ* polymerase chain reaction in cell preparations and tissue sections. Detection of viral DNA gene rearrangements and chromosomal translocations. Histochemistry *99*:151-162.
40. **Long, A.A., P. Komminoth and H.J. Wolfe.** 1992. Detection of HIV provirus by *in situ* polymerase chain reaction (letter). N. Engl. J. Med. *327*:1529-1530.
41. **Lundberg, K.S., D.D. Shoemaker, M.W. Adams, J.M. Short, J.A. Sorge and E.J. Mathur.** 1991. High fidelity amplification using a thermostable DNA polymerase isolated from Pyrococcus furiosus. Gene *108*:1-6.
42. **Mullis, K.B., F. Faloona, S.J. Scharf, R.K. Saiki, G.T. Horn and H.A. Erlick.** 1986. Specific enzymatic amplification of DNA in vitro: The polymerase chain reaction. Cold Spring Harbor Symp. Quant. Biol. *51*:263-273.
43. **Nuovo, G.J.** 1989. Buffered formalin is the superior fixative for the detection of human

papillomavirus DNA by *in situ* hybridization analysis. Am. J. Pathol. *134*:837-842.

44. **Nuovo, G.J.** 1992. *In Situ* Hybridization. Protocols and Applications. Raven Press, New York.

45. **Nuovo, G.J., J. Becker, M. Margiotta, P. MacConnell, S. Comite and H. Hochman.** 1992. Histological distribution of polymerase chain reaction-amplified human papillomavirus 6 and 11 DNA in penile lesions. Am. J. Surg. Pathol. *16*:269-275.

46. **Nuovo, G.J., A. Forde, P. MacConnell and R. Fahrenwald.** 1993. *In situ* detection of PCR-amplified HIV-1 nucleic acids and tumor necrosis factor cDNA in cervical tissues. Am. J. Pathol. *143*:40-48.

47. **Nuovo, G.J., F. Gallery, R. Hom, P. MacConnell and W. Bloch.** 1993. Importance of different variables for enhancing *in situ* detection of PCR-amplified DNA. PCR Methods Appl. *2*:305-312.

48. **Nuovo, G.J., F. Gallery and P. MacConnell.** 1992. Detection of amplified HPV 6 and 11 DNA in vulvar lesions by hot start PCR *in situ* hybridization. Mod. Pathol. *5*:444-448.

49. **Nuovo, G.J., F. Gallery, P. MacConnell, J. Becker and W. Bloch.** 1991. An improved technique for the *in situ* detection of DNA after polymerase chain reaction amplification. Am. J. Pathol. *139*:1239-1244.

50. **Nuovo, G.J., G.A. Gorgone, P. MacConnell, M. Margiotta and P.D. Gorevic.** 1992. *In situ* localization of PCR-amplified human and viral cDNAs. PCR Methods Appl. *2*:117-123.

51. **Nuovo, G.J., K. Lidonnici, P. MacConnell and B. Lane.** 1993. Intracellular localization of polymerase chain reaction (PCR)—amplified hepatitis C cDNA. Am. J. Surg. Pathol. *17*:683-690.

52. **Nuovo, G.J., P. MacConnell, A. Forde and P. Delvenne.** 1991. Detection of human papillomavirus DNA in formalin-fixed tissues by *in situ* hybridization after amplification by polymerase chain reaction. Am. J. Pathol. *139*:847-854.

53. **Nuovo, G.J., M. Margiotta, P. MacConnell and J. Becker.** 1992. Rapid *in situ* detection of PCR-amplified HIV-1 DNA. Diagn. Mol. Pathol. *1*:98-102.

54. **Nuovo, G.J. and S.J. Silverstein.** 1988. Comparison of formalin, buffered formalin, and Bouin's fixation on the detection of human papillomavirus DNA from genital lesions. Lab. Invest. *59*:720-724.

55. **Nuovo, M.A., G.J. Nuovo, P. MacConnell, A. Forde and G.C. Steiner.** 1992. *In situ* analysis of Paget's disease of bone for measles-specific PCR-amplified cDNA. Diagn. Mol. Pathol. *1*:256-265.

56. **Patel, V.G., A. Shum-Siu, B.W. Heniford, T.J. Wieman and F.J. Hendler.** Detection of epidermal growth factor receptor mRNA in tissue sections from biopsy specimens using *in situ* polymerase chain reaction. Am. J. Pathol. *144*:7-14.

57. **Patterson, B.K., M. Till, P. Otto and C. Goolsby, M.R. Furtado, L.J. McBride and S.M. Wolinsky.** 1993. Detection of HIV-1 DNA and messenger RNA in individual cells by PCR-driven *in situ* hybridization and flow cytometry. Science *260*:976-979.

58. **Ray, R., P. Komminoth, M. Machado and H.J. Wolfe.** 1991. Combined polymerase chain reaction and *in situ* hybridization for the detection of single copy genes and viral genomic sequences in intact cells. Mod. Pathol. *4*:124A.

59. **Saiki, R.K.** 1989. The design and optimization of the PCR, p. 17-22. *In* H.A. Erlich (Ed.). PCR Technology. Stockton Press, New York.

60. **Sanger, F., S. Nicklen and A.R. Coulson.** 1977. DNA sequencing with chain-terminating inhibitors. Proc. Natl. Acad. Sci. USA *74*:5463-5468.

61. **Smith-Norowitz, T., E. Zevallos, V. Anderson and H.G. Durkin.** 1994. A modified *in situ* PCR-hybridization procedure. *1*:84.

62. **Spann, W., K. Pachmann, H. Zabnienska, A. Pielmeier and B. Emmerich.** 1991. *In situ* amplification of single copy gene segments in individual cells by the polymerase chain reaction. Infection *19*:242-244.

63. **Stanta, G. and C. Schneider.** 1991. RNA extracted from paraffin-embedded human tissues is amenable to analysis by PCR amplification. BioTechniques *11*:3.

64. **Staskus, K.A., L. Couch, P. Bitterman, E.F. Retzel, M. Zupancic, J. List and A.T. Haase.** 1991. *In situ* amplification of visna virus DNA in tissue sections reveals a reservoir of latently infected cells. Microb. Pathog. *11*:67-76.

65. **Stork, P., M. Loda, S. Bosari, B. Wiley, K. Poppenhusen and H.J. Wolfe.** 1992. Detection of K-ras mutations in pancreatic and hepatic neoplasms by non-isotopic mismatched polymerase chain reaction. Oncogene *6*:857-862.

66. **Wilkinson, D.G. (Ed.).** 1993. *In situ* Hybridization: A Practical Approach. IRL Press at Oxford University Press, New York.

67. **Yap, E.P.H. and J.O'D. McGee.** 1991. Slide PCR: DNA amplification from cell samples on microscopic glass slides. Nucleic Acids Res. *19*:15.
68. **Yin, J., M.G. Kaplitt and D.W. Pfaff.** 1994. *In situ* PCR and *in vivo* detection of foreign gene expression in rat brain. Cell Vision *1*:58-59.
69. **Zehbe, I., G.W. Hacker, E. Rylander, J. Sällström and E. Wilander.** 1992. Detection of single HPV copies in SiHa cells by *in situ* polymerase chain reaction (*in situ* PCR) combined with immunoperoxidase and immunogold-silver staining (IGSS) techniques. Anticancer Res. *12*:2165-2168.
70. **Zehbe, I., G.W. Hacker, J.F. Sällström, W.H. Muss, C. Hauser-Kronberger, E. Rylander and E. Wilander.** 1994. Polymerase chain reaction *in situ* hybridization (PISH) and *in situ* self-sustained sequence replication based amplification (*in situ* 3SR). Cell Vision *1*:46-47.
71. **Zehbe, I., G.W. Hacker, J.F. Sällström, E. Rylander and E. Wilander.** 1994. Self-sustained sequence replication based amplification (3SR) for the *in situ* detection of mRNA in cultured cells. Cell Vision *1*:20-24.
72. **Zebhe, I., J.F. Sällström, G.W. Hacker, C. Hauser-Kronberger, E. Rylander and E. Wilander.** 1994. Indirect and direct *in situ* PCR for the detection of human papillomavirus. An evaluation of two methods and a double staining technique. Cell Vision *1*:163-167.
73. **Zebhe, I., J.F. Sällström, G.W. Hacker, E. Rylander, A. Strand, A.-H. Graf and E. Wilander.** 1994. Polymerase chain reaction (PCR) *in situ* hybridization: Detection of human papillomavirus (HPV) DNA in SiHa cell monolayers. *In* J. Gu and G.W. Hacker (Eds.), Modern Methods in Analytical Morphology, Chapter 18. Plenum Publishing Corp. (In press).
74. **Zevallos, E., V. Anderson, E. Bard, T. Smith, M. Nowakoski and J. Gu.** 1994. Reverse transcribed *in situ* RT-ISPCR of retinoblastoma mRNA. Cell Vision *1*:88.
75. **Zevallos, E., V. Anderson, M. Nowakoski, E. Bard and J. Gu.** 1994. Detection of HIV-1 RNA in lung and bronchioloalveolar lavage cells by reverse transcribed *in situ* PCR. Cell Vision *1*:87.
76. **Zevallos, E., V. Anderson, M. Nowakoski, E. Bard and J. Gu.** 1994. Detection of HIV-1 sequence in lung and bronchioloalveolar lavage by *in situ* PCR. Cell Vision *1*:87.
77. **Zevallos, E., V. Anderson, M. Nowakoski, E. Bard and J. Gu.** 1994. Detection of human retinoblastoma gene expression by rTth-driven reverse transcribed *in situ* PCR. Cell Vision *1*:88.
78. **Zevallos, E., E. Bard, V. Anderson, N. Carson and J. Gu.** 1994. HIV *in situ* PCR and reverse transcribed *in situ* PCR. Cell Vision *1*:52-54.
79. **Zevallos, E., E. Bard, V. Anderson and J. Gu.** 1994. An *in situ* (ISPCR) study of HIV-1 infection of lymphoid tissues and peripheral lymphocytes. Cell Vision *1*:87.
80. **Zevallos, E., E. Bard, V. Anderson and J. Gu.** 1994. Detection of HIV-1 sequences in placentas of HIV-infected mothers by *in situ* PCR. Cell Vision *1*:116-121.

Address correspondence to: Jiang Gu, M.D., Ph.D., Chairman, Scientific Affairs, Deborah Research Institute, Trenton-Pine Mill Road, Browns Mills, NJ 08015-1799.

In Situ PCR: General Methodology and Recent Advances

Aidan A. Long[1] and Paul Komminoth[2]

[1]Department of Medicine, Allergy Division, New England Medical Center Hospitals and Tufts University, Boston, MA, USA and [2]Department of Pathology, Division of Cell and Molecular Pathology, University of Zürich, Zürich, Switzerland

SUMMARY

In situ *PCR is a new molecular technique that combines the extreme sensitivity of PCR with the cellular localization provided by* in situ *hybridization (ISH). It is based on the principles of amplification of specific gene sequences within intact cells or tissue sections in order to increase copy numbers to levels detectable by ISH or immunohistochemistry.*

Methods of in situ PCR have been independently developed in several laboratories and include a number of different techniques, which are not equally applicable to all types of samples. In this chapter we examine the principles of in situ *PCR and compare the different protocols. Our own experience suggests that* in situ *PCR appears to be most effective for the detection of DNA in single-cell preparations with controlled fixation and pretreatment, although the quantification of results remains problematic.* In situ *PCR works less efficiently in archival tissue sections due to poor quality of nucleic acids and poor retention of PCR products. Emphasis is placed on the absolute requirement for meaningful controls to allow accurate interpretation of data. Artifacts caused by diffusion and extracellular generation of PCR products are significant problems, which can potentially lead to false-positive results. Direct* in situ *PCR yields less specific results than indirect* in situ *PCR and requires additional controls, such as omission of primers in the reaction mixture to detect artifacts. Nonspecific results seen in direct* in situ *PCR are due to incorporation of labeled nucleotides into fragmented genomic DNA undergoing repair by the DNA polymerase or into PCR products, generated by endogenous DNA or cDNA fragments ("endogenous priming").*

In situ *PCR, most likely the indirect approach, will emerge with an important role in specialized diagnostic laboratories in the evaluation of viral diseases, hematological and other malignancies with unique genetic markers.*

INTRODUCTION

The polymerase chain reaction (PCR) and *in situ* hybridization (ISH) are established techniques in the study of gene structure and expression. With ISH, the localization of specific nucleic acid sequences within individual cells is achieved, but the application of this technique is occasionally limited by low sensitivity of detection. PCR, on the other hand, is a technique of extremely high sensitivity with the potential to amplify rare or single copy gene sequences to levels easily detectable by gel electrophoresis and Southern blot hybridization. Since conventional PCR generally requires cell or tissue destruction to isolate nucleic acids, one can neither correlate results of gene detection or expression with histopathological features nor estimate the frequency of cells that contain the target sequence.

Several recent studies have described a methodology that combines the high sensitivity of PCR with the cytological localization of sequences provided by ISH. The combination of these techniques has been variously termed "*in situ* PCR" (33), "PCR *in situ*" (2), "PCR *in situ* hybridization" (24), "in cell PCR" (1) or "PCR-driven ISH" (32). For the detection of low copy RNA sequences, the addition of an intracellular reverse transcription (RT) step to generate cDNA from RNA templates prior to *in situ* PCR has been used. This modification of *in situ* PCR has been termed "*in situ* RT-PCR" (17) or less precisely "RT *in situ* PCR" (24) or "*in situ* cDNA PCR" (3).

PRINCIPLES AND METHODS OF *IN SITU* PCR

In situ PCR represents an attempt to combine the techniques of PCR and ISH through amplification of specific nucleic acid sequences inside single cells to make them more easily detectable by ISH or immunohistochemistry (see Figure 1). Experimental protocols for successful *in situ* PCR have been independently developed by several groups, and a majority of these protocols share important key characteristics described below.

Sample Preparation

Prior to *in situ* PCR, cells or tissue samples need to be fixed and permeabilized. These pretreatments have the dual purposes of preserving cellular morphology and permitting access of the PCR reagents to the intracellular nucleic acid sequences to be amplified. It has now been established that not all fixatives appear equally

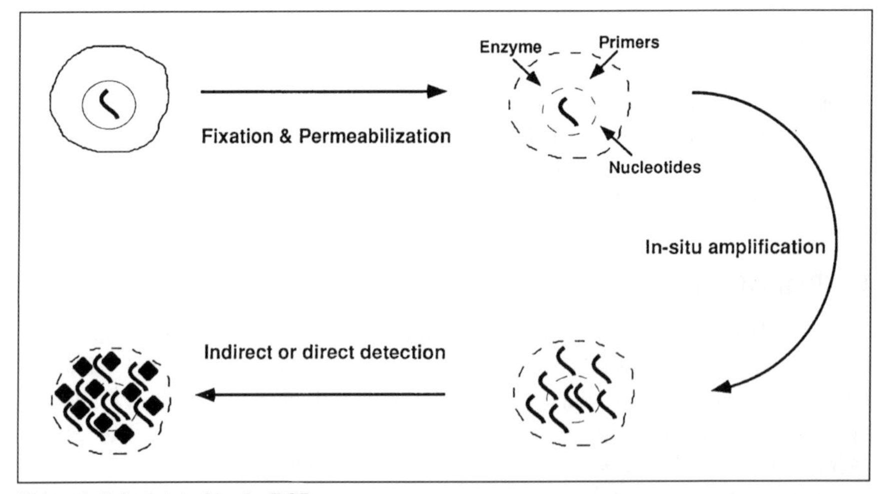

Figure 1. Principles of *in situ* PCR.

permissive to *in situ* PCR. Successful *in situ* amplification has been reported in samples fixed in 1%–4% paraformaldehyde, 10% buffered formaldehyde solution, alcohol or mixtures of alcohol and acetic acid (2,4,12,19,22). In general terms, longer fixation times seem to necessitate some additional digestion with proteases or detergents, presumably to increase accessibility of target nucleic acid sequences (2,18,19,24). Employment of extensive digestion steps can result in substantial destruction of the cell architecture, and prolonged permeabilization of formaldehyde-fixed samples increases risk of false positives in experiments using direct *in situ* PCR (see later), a finding probably related to the increased removal of cross-linked histones by proteases with consequent unmasking of DNA nicks (30). Different approaches are needed for cell preparations and tissue sections and, in our experience, the cleanest results from *in situ* PCR performed in cells in suspension are achieved if no proteolytic digestion is employed (19). It is suggested that for new materials, optimal conditions be established by performing lysis experiments after attempted *in situ* PCR and analyzing the products of *in situ* PCR by gel electrophoresis to document whether or not intracellular amplification has been successful.

In Situ Amplification

The nature of the cellular material containing the target sequences to be amplified is an important variable in *in situ* PCR. *In situ* PCR amplification has been successfully achieved in cells held in suspension in microcentrifuge tubes and also on cytocentrifuge preparations or tissue sections directly under a coverslip on glass slides.

In situ PCR with cells in suspension, first described by Haase et al. (12), is an approach that has been employed by several groups. It is performed with fixed cells suspended in a PCR mixture in a microcentrifuge tube in a conventional thermal cycler. After PCR, the cells are recovered and cytocentrifuged onto glass slides. Visualization of intracellular PCR products is then achieved by either ISH or immunohistochemistry.

Our results to date indicate that this approach seems to provide optimal physical conditions for thermal cycling and yields maximal cellular target sequence amplification (19,22). Nucleic acids are probably better preserved in single cell preparations, and the fixed intact cells may be thought of as "amplification sacks" with semi-permeable membranes. The membranes permit the primers, nucleotides and the DNA polymerase enzyme to pass into the cell and nucleus, yet seem to sufficiently retard the outward diffusion of PCR products to allow for their detection *in situ*.

For *in situ* PCR directly performed on glass slides, the cellular material is overlayed with PCR mixture under a coverslip. The coverslip is sealed or covered with mineral oil to prevent evaporation of the PCR mixture. Thermal cycling is achieved by placing the slides either directly on top of the heating block of a conventional thermocycler or by using specially designed equipment (e.g., OmniSlide™ system, Hybaid, Teddington Middlesex, UK; PTC-100-12MS Programmable Thermal Controller, MJ Research, Watertown, MA, USA; GeneAmp™ *in situ* PCR system

1000, Perkin-Elmer, Norwalk, CT, USA) or thermal cycling ovens (e.g., Bio-Oven™, Biotherm Corp, Fairfax, VA, USA).

When compared to solution-phase PCR, the efficiency of *in situ* DNA amplification is very low. Amplification has been estimated to be approximately 50-fold after 30 cycles in suspended cells and is lower again in cytospin preparations and in tissue sections (19). Several factors probably contribute to the apparent decreased efficiency of *in situ* PCR in tissue sections. For example, target sequences are possibly fragmented during the fixation, embedding and storage processes; microtomy leaves many cells without intact cellular or nuclear membranes such that PCR products are probably poorly retained *in situ* after synthesis.

Detection of Intracellular PCR Products

Detection of intracellular PCR products has been achieved by two entirely different methods: indirectly by ISH (indirect *in situ* PCR) or directly through detection of labeled (digoxigenin-11-dUTP, flourescein-dUTP, ^{3}H-CTP) nucleotides that have been incorporated into PCR products during thermal cycling (direct *in situ* PCR).

By using probes that specifically recognize only the amplified sequences, the indirect method provides maximum specificity in detection of intracellular PCR products and is the approach that most investigators have used. The detection step is carried out in a manner identical to conventional ISH.

Direct *in situ* PCR, by obviating the need for ISH, was initially heralded as a rapid alternative to indirect *in situ* PCR (13,24,26,30,31,37) but has been shown to yield far less reliable results than indirect *in situ* PCR (17,20,22). Indeed, a concensus of opinion at a recent workshop focusing on *in situ* PCR stated that, since it has not been possible to adequately circumvent nonspecific artifact in direct ISPCR, the direct approach, as currently formulated, has no meaningful role as a technology for the intracellular localization of nucleic acid sequences (2nd International Workshop on Modern Methods in Analytical Morphology, April 4–5, 1994, Orlando, Florida. Proceedings published in *Cell Vision—Journal of Analytical Morphology*, Vol. 1, No. 1). This does not preclude modifications of the method, which may improve specificity of direct *in situ* PCR.

CONTROLS

The importance of appropriate and adequate controls cannot be overemphasized. In the relatively short history of *in situ* PCR, failure to adequately control for artifacts has already resulted in significant error in interpretation of results. Tissue controls, internal controls, reagent controls and technical controls all have a place. Controls should be both qualitative, to exclude false-positive and false-negative results, as well as quantitative. A tabulation of appropriate controls to consider in designing experiments is included in Table 1.

In our experience, omission of primers in the PCR mixture is the single most important control to detect artifacts in direct *in situ* PCR experiments, both for DNA and cDNA detection (*in situ* RT-PCR) (18).

Table 1. Controls Required for *In Situ* PCR Experiments

General Controls

Control	Purpose
- Use of known positive and negative control samples	Control of specificity and sensitivity of method used
- Solution-phase PCR on extracted DNA/RNA of actual test samples	Detection of false negatives, control of sensitivity
- Amplification of endogenous control DNA sequences	Detection of false negatives, control of DNA/RNA quality
- Omission of primary antibody in immunohistochemical detection	Detection of endogenous enzyme activity

Additional controls

Method	Control(s)	Purpose
Cells in suspension	- Lysis of cells after *in situ* PCR and analysis of PCR products by gel electrophoresis, Southern blot hybridization (and sequencing)	Specificity control of PCR products
Indirect *in situ* PCR	- Omission of DNA polymerase	Detection of nonspecific probe and antibody sticking
	- Use of irrelevant probes for *in situ* hybridization	Control of *in situ* hybridization specificity
Direct *in situ* PCR	- Omission of primers	Detection of artifacts related to DNA repair and endogenous priming
	- Omission of DNA polymerase	Detection of nonspecific probe and antibody sticking
In situ RT-PCR	- Omission of reverse transcriptase step - RNase pretreatment of samples	Detection of mis-priming and amplification of endogenous DNA
Quantitative *in situ* PCR	- *In situ* PCR in mixtures of known positive and negative cells of different proportions - Identification of different cell types by immunohistochemistry	Control of specificity and sensitivity of method used

VARIABLES IN *IN SITU* PCR

There is no generally applicable single *in situ* PCR protocol, and significant experimental differences in experimental methods are found among the published protocols. Accurate interpretation and meaningful comparisons of results obtained with different published *in situ* PCR protocols require careful consideration of both technical details and control experiments. Important variables appear to include the type of starting material (suspended cells, cytospins, tissue sections), the type and copy number of target sequence (viral or genomic DNA or RNA), the DNA ampli-

Table 2. Applications of *In Situ* PCR

Foreign DNA/RNA		
Detection	**Method**	**Reference**
Lentivirus DNA	Indirect *in situ* PCR, cells in suspension	Haase et al. 1990
	Indirect *in situ* PCR, tissue sections	Staskus et al. 1991
Human papilloma virus DNA	Indirect *in situ* PCR, tissue sections	Nuovo et al. 1991, 1992
	Direct *in situ* PCR, cyto-centrifuge preparations	Zehbe et al. 1992
Human immundeficiency virus DNA	Direct *in situ* PCR, cyto-centrifuge preparations	Spann et al. 1991
	Direct *in situ* PCR, cyto-centrifuge preparations	Nuovo et al. 1992a
	Direct *in situ* PCR, cyto-centrifuge preparations	Bagasra et al. 1992, 1993
	Indirect *in situ* PCR, tissue sections	Embretson et al. 1993a,b
Human immundeficiency virus mRNA	Indirect *in situ* RT-PCR, cells in suspension	Patterson et al. 1993
Mouse mammary tumor virus DNA	Indirect *in situ* PCR, tissue sections	Chiu et al. 1992
Hepatitis C virus RNA	Direct *in situ* RT-PCR, tissue sections	Nuovo et al. 1993
	Indirect *in situ* RT-PCR, tissue sections	Komminoth et al. 1994
Cytomegalovirus DNA	Indirect *in situ* PCR, cells in suspension	Komminoth et al. 1992
	Indirect *in situ* PCR, tissue sections	Long et al. 1993
Hepatitis B virus DNA	Indirect *in situ* PCR, tissue sections	Long et al. 1993
Herpes simplex virus 2 DNA	Direct *in situ* PCR, tissue sections	Gressens et al. 1994
JC virus RNA	Direct *in situ* RT-PCR, tissue sections	Kelleher et al. 1994
Pneumocystis carinii DNA	Direct *in situ* PCR, tissue sections	Tsongalis et al. 1994

Endogenous DNA/RNA		
Detection	**Method**	**Reference**
Gene rearrangements, DNA	Indirect *in situ* PCR, cells in suspension	Komminoth et al. 1992a
Gene rearrangments, mRNA	Indirect *in situ* RT-PCR, cells in suspension	Embleton et al. 1992
Chromosomal translocations, DNA	Indirect *in situ* PCR, cells in suspension	Long et al. 1993
Chromosome mapping, DNA	Direct *in situ* PCR, metaphase preparations	Gosden et al. 1993
Nerve growth factor mRNA	Direct *in situ* RT-PCR, tissue sections	Staecker et al. 1993
Granzyme A, perforin mRNA	Direct *in situ* RT-PCR, tissue sections	Chen et al. 1993
Epidermal growth factor receptor mRNA	Direct *in situ* RT-PCR, tissue sections	Patel et al. 1994

fication method (i.e., with single or multiple primer pairs), the detection system (direct or indirect *in situ* PCR) and the use of adequate control experiments (18,23).

Despite its conceptual simplicity, therefore, the precise mechanisms underlying successful *in situ* PCR remain speculative. At this writing we do not completely understand how different methods of cellular processing, fixation and digestion with proteolytic enzymes affect accessibility of target DNA or RNA to the PCR components. We do not have insight into the reasons for the apparent reduced efficiency of PCR in the intracellular location. Nor has it been adequately determined to what extent improved efficiency can be achieved by adjusting concentrations of the various components of the PCR mixture (e.g., polymerase, Mg^{2+}, bovine serum albumin). We do not have a clear picture of what the determinants of successful retention of PCR products are at the site of amplification. It is possible that some of these variables relate to the physical characteristics of the compartment in which the reaction is taking place, such as the high surface area contributed to by membranes and organelles available to adsorb reagents. They may be related to protein binding of the target nucleic acids, rendering them unavailable to the PCR reagents. Improved product retention might simply be a matter of decreased diffusibility related to size or bulk of the PCR products; or it may result from the formation of a lattice or scaffolding between combinations of long and short products and cytoskeletal components holding the amplified DNA *in situ*. We are in the developmental stages of this new technology and have much to learn.

APPLICATIONS OF *IN SITU* PCR

In situ PCR has a number of potential research and diagnostic applications (Table 2). To date, many groups have reported successful *in situ* PCR-based detection of specifically amplified single copy nucleic acid sequences in single cells and low copy DNA sequences in tissue sections.

Most of the studies have focused on the detection of viral and proviral (foreign) nucleic acid sequences (e.g., Visna-virus, HIV, HPV, MMTV, CMV, HBV, HCV) (1,2,4–6,9,10,12,15,17,19,22,23,25–29,32,35,36,40).

In addition, *in situ* PCR has also been applied to study endogenous DNA sequences, including human single copy genes, rearranged cellular genes and chromosomal translocations (19,22) and to map low copy number genomic sequences in metaphase chromosomes (9). An elegant application of *in situ* PCR has recently been realized in transgenic animals, namely following the anatomic localization of the introduced gene (39). Localization of mutated genes including proto-oncogenes will likely be amenable to study by this technology.

Recently, successful amplification and detection of low copy mRNA (3,5,31,36) and viral RNA (15,17,27,32) sequences has also been reported.

PITFALLS OF *IN SITU* PCR

A cursory reading of the literature on *in situ* PCR might suggest that this technique is both straightforward and problem-free, but both false-positive and false-negative results have been seen in *in situ* PCR (17,20,21,34).

Table 3. Nonspecific Pathways of *In Situ* PCR

False Positives	
Method	**Artifact(s)**
(Direct) *in situ* (RT) PCR	Mis-priming
in situ (RT) PCR in cells in suspension	"Diffusion" artifacts
Direct *in situ* (RT) PCR	"DNA repair" artifacts
	"Endogenous priming"
False Negatives	
Method	**Artifact(s)**
in situ (RT) PCR on glass slides	Reduced amplification efficiency
in situ (RT) PCR on glass slides	Loss of PCR products

False-Positive Results

Nonspecific pathways leading to false-positive results include diffusion artifacts, DNA-repair artifacts and mispriming (Table 3). "Diffusion artifacts" have been reported as a significant problem in which PCR products and/or template DNA can leak out of positive cells and serve as templates for extracellular amplification. These amplified DNA sequences have the potential to adhere to the surface of adjacent template negative cells or perhaps to even diffuse into them, resulting in false-positive signals (19,22). In agreement with Haase et al. (12), we have observed more problems related to diffusion artifacts when short DNA sequences were amplified by *in situ* PCR. Many creative approaches have been employed to minimize the impact of this phenomenon, and it is now clear that diffusion artifacts can be significantly reduced by optimal fixation and permeabilization, reduction of PCR cycle numbers, generation of longer or more complex (e.g., overlapping) PCR products or by incorporation of substituted (e.g., biotinylated) nucleotides to generate bulkier and therefore less diffusible PCR products (19,22).

In direct *in situ* PCR experiments false-positive results mainly result from nonspecific incorporation of labeled nucleotides. The nonspecific incorporation occurs through fragmented endogenous DNA undergoing "repair" by the DNA polymerase or by priming of nonspecific PCR by intracellular cDNA or DNA fragments during thermal cycling (4,17,22,34). These artifacts typically exhibit nuclear signals and are most marked in work on tissue sections where nucleic acid sequences are damaged during processing. They are particularly evident in apoptotic cells where DNA fragmentation is a key feature (8,34,38). It has not yet been possible to circumvent this nonspecific pathway of direct *in situ* PCR. Despite controlled fixation and protease digestion, hot-start modifications of *in situ* PCR and DNase pretreatments to reduce endogenous DNA amplification in *in situ* RT-PCR experiments, direct *in situ* PCR still yields false-positive results marked enough to significantly interfere with accurate detection of "specific" signal. The artifacts can be somewhat reduced by using an exonuclease-free DNA polymerase (17) or repairing DNA nicks by treatment with T4 DNA ligase (16) or by initial thermal cycling using unlabeled or dideoxy nucleotides, but they do not disappear completely. Mispriming can also result in nonspecific single-stranded PCR products primed by endogenous DNA, cDNA fragments ("endogenous priming") or by PCR primers used under sub-optimal stringencies. The phenomena of "DNA repair" and "endogenous priming" are both DNA polymerase and cycling dependent. They occur despite omission of primers and hot-start *in situ* PCR and the use of exonuclease-free DNA polymerases. Factors that influence mispriming include the specificity of the oligonucleotide primers, the pH of the reaction and the annealing temperature used during thermal cycling. Accordingly, great attention needs to be paid to primer design to maximize specificity, minimize the risk of primer-dimer formation and to assure the compatibility of annealing temperatures between specific members of primer pairs. Many computer software programs are now available to facilitate the choice of optimal primers. In general, Mg^{2+} concentrations need to be higher for *in situ* PCR than for solution-phase PCR. A series of experiments to optimize this is strongly recommended in any new application.

False-Negative Results

False-negative or inconsistent results most often occur in *in situ* PCR experiments performed on glass slides (17,20,22). Possible explanations for this finding are the reduced amplification efficiency on glass slides due to mechanical factors, such as poor thermal conduction, uneven convection patterns in the PCR mixture under a coverslip or possible adsorption of DNA polymerase to glass. Alternatively, loss of PCR products during washing procedures or tissue-related factors including inhibitors of the DNA polymerase and poor quality of target sequences may be important. We and others have only been able to amplify DNA sequences in tissue sections using multiple primer pairs and relatively long DNA probes or cocktails of oligonucleotide probes (4–6,17,20,22). While this may be related to the poor amplification efficiency in tissue sections, it is possible that the positive results seen after *in situ* PCR with multiple primer pairs are due to the formation of a "scaffolding" of overlapping PCR products. This helps to anchor the PCR products in place and make them less susceptible to being washed away during the detection steps (22).

CONCLUSIONS

In situ PCR is a new and exciting technology that is already providing a mechanism to gain insights into disease pathogenesis (6,7,11). As with other emerging technologies, its true strengths and weaknesses are becoming clearer with time. That PCR is a highly sensitive, specific, reliable and reproducible procedure is now beyond question. That this method of nucleic acid amplification can be readily transferred to an intracellular location is also becoming certain.

In situ PCR encompasses a number of different techniques not all of which are equally applicable to different starting materials. It appears to be most effective for low copy DNA detection in single-cell preparations after controlled fixation and pretreatment. However, the exact quantification of results remains problematic (22). *In situ* PCR on tissue sections appears to require the use of multiple primer pairs for target sequence amplification and genomic probes or cocktails of oligonucleotide probes for PCR product detection. Direct *in situ* PCR yields significantly less specific results than indirect *in situ* PCR and is not applicable on tissue sections.

In situ PCR has important potential in research and diagnostics. It would be most useful in situations where ISH fails due to low copy numbers and PCR positivity should be correlated to histopathological features. Possible applications of *in situ* PCR include the localization, distribution and quantification of cells with latent viral infection and specific chromosomal translocations or gene rearrangements. However, clinical application of this technology must await resolution of its current limitations (14).

REFERENCES

1. **Bagasra, O., S. Hauptman, H. Lischner, M. Sachs and R. Pomerantz.** 1992. Detection of human immunodeficiency virus type 1 provirus in mononuclear cells by in situ polymerase chain reaction. N. Engl. J. Med. *326*:1385-1391.
2. **Bagasra, O., T. Seshamma and R. Pomerantz.** 1993. Polymerase chain reaction in situ: Intracellular amplification and detection of HIV-1 proviral DNA and other specific genes. J. Immunol. Methods *158*:131-145.
3. **Chen, R.H and S.V. Fuggle.** 1993. In situ cDNA polymerase chain reaction. A novel technique for detecting mRNA expression. Am. J. Pathol. *143*:1527-1534.
4. **Chiu, K.-P., S. Cohen, D. Morris and G. Jordan.** 1992. Intracellular amplification of proviral DNA in tissue sections using the polymerase chain reaction. J. Histochem. Cytochem. *40*:333-341.
5. **Embleton, M., G. Gorochov, P. Jones and G. Winter.** 1992. In-cell PCR from mRNA: Amplifying and linking the rearranged immunoglobulin heavy and light chain V-genes within single cells. Nucleic Acids Res. *20*:3831-3837.
6. **Embretson, J., M. Zupancic, J. Beneke, M. Till, S. Wolinsky, J. Ribas, A. Burke and A. Haase.** 1993. Analysis of human immunodeficiency virus-infected tissues by amplification and in situ hybridization reveals latent and permissive infections at single-cell resolution. Proc. Natl. Acad. Sci. USA *90*:357-361.
7. **Embretson, J., M. Zupancic, J. Ribas, A. Burke, P. Racz, K. Tenner-Racz and A. Haase.** 1993. Massive covert infection of helper T lymphocytes and macrophages by HIV during the incubation period of AIDS. Nature *362*:359-362.
8. **Gold, R., M. Schmied, G. Rothe, H. Zischler, H. Breitschopf, H. Wekerle and H. Lassmann.** 1993. Detection of DNA fragmentation in apoptosis: Application of in situ nick translation to cell culture systems and tissue sections. J. Histochem. Cytochem. *41*:1023-1030.
9. **Gosden, J. and D. Hanratty.** 1993. PCR *in situ*: A rapid alternative to *in situ* hybridization for mapping short, low copy number sequences without isotopes. BioTechniques *15*:78-80.
10. **Gressens, P. and J.R. Martin.** 1994. HSV-2 DNA persistence in astrocytes of the trigeminal root entry zone: Double labeling by in situ PCR and immunohistochemistry. J. Neuropathol. Exp. Neurol. *53*:127-135.
11. **Haase, A.T.** 1987. Analysis of viral infections by in situ hybridization, p. 197-219. *In* K.L. Valentino, J.H. Eberwine and J.D. Barchas (Eds.), In Situ Hybridization. Applications to Neurobiology. Oxford University Press, New York.
12. **Haase, A.T., E.F. Retzel and K.A. Staskus.** 1990. Amplification and detection of lentiviral DNA inside cells. Proc. Natl. Acad. Sci. USA *87*:4971-4975.
13. **Heniford, B.W., S.A. Shum, M. Leonberger and F.J. Hendler.** 1993. Variation in cellular EGF receptor mRNA expression demonstrated by in situ reverse transcriptase polymerase chain reaction. Nucleic Acids Res. *21*:3159-3166.
14. **Höfler, H.** 1993. In situ polymerase chain reaction: Toy or tool? Histochemistry *99*:103-104.
15. **Kelleher, M.B., D. Galutira, T.D. Duggan and G.J. Nuovo.** 1994. Progressive multifocal leukoencephalopathy in a patient with Alzheimer's disease. Diagn. Mol. Pathol. *3*:105-113.

16. **Koch, J., J. Hindkjaer, J. Mogensen, S. Kolvraa and L. Bolund.** 1991. An improved method for chromosome-specific labeling of alpha satellite DNA in situ by using denatured double-stranded DNA probes as primers in a primed in situ labeling (PRINS) procedure. Genet. Anal. Tech. Appl. *8*:171-178.
17. **Komminoth, P., V. Adams, A.A. Long, J. Roth, P. Saremaslani, R. Flury, M. Schmid and P.U. Heitz.** 1994. Evaluation of methods for hepatitis C virus (HCV) detection in liver biopsies: Comparison of histology, immunohistochemistry, in-situ hybridization, reverse transcriptase (RT) PCR and in-situ RT PCR. Pathol. Res. Pract. (In press).
18. **Komminoth, P. and A. Long.** 1993. In-situ polymerase chain reaction. An overview of methods, applications and limitations of a new molecular technique. Virchows Arch. [B] *64*:67-73.
19. **Komminoth, P., A. Long, R. Ray and H. Wolfe.** 1992. In situ polymerase chain reaction detection of viral DNA, single copy genes and gene rearrangements in cell suspensions and cytospins. Diagn. Mol. Pathol. *1*:85-97.
20. **Komminoth, P., A. Long and H. Wolfe.** 1992. Comparison of in-situ polymerase chain reaction (in-situ PCR), in-situ hybridization (ISH) and polymerase chain reaction (PCR) for the detection of viral infection in fixed tissue. Patologia Suppl. *25*:253.
21. **Komminoth, P. and A.A. Long.** 1994. Non-specific pathways of in-situ PCR. Am. J. Pathol. (In press).
22. **Long, A., P. Komminoth and H. Wolfe.** 1993. Comparison of indirect and direct in-situ polymerase chain reaction in cell preparations and tissue sections. Detection of viral DNA, gene rearrangements and chromosomal translocations. Histochemistry *99*:151-162.
23. **Long, A., P. Komminoth and H. Wolfe.** 1992. Detection of HIV provirus by in situ polymerase chain reaction. N. Engl. J. Med. *327*:1529.
24. **Nuovo, G.** 1992. PCR *In Situ* Hybridization. G. Nuovo (Ed.). Raven Press, New York.
25. **Nuovo, G., J. Becker, M. Margiotta, P. MacConnell, S. Comite and H. Hochman.** 1992. Histological distribution of polymerase chain reaction-amplified human papillomavirus 6 and 11 DNA in penile lesions. Am. J. Surg. Pathol. *16*:269-275.
26. **Nuovo, G., F. Gallery, P. MacConnell, J. Becker and W. Bloch.** 1991. An improved technique for the in situ detection of DNA after polymerase chain reaction amplification. Am. J. Pathol. *139*:1239-1244.
27. **Nuovo, G., K. Lidonnici, P. MacConnell and B. Lane.** 1993. Intracellular localization of polymerase chain reaction (PCR)-amplified hepatitis C cDNA. Am. J. Surg. Pathol. *17*:683-690.
28. **Nuovo, G., P. MacConnell, A. Forde and P. Delvenne.** 1991. Detection of human papillomavirus DNA in formalin-fixed tissues by in situ hybridization after amplification by polymerase chain reaction. Am. J. Pathol. *139*:847-854.
29. **Nuovo, G., M. Margiotta, P. MacConnell and J. Becker.** 1992. Rapid in situ detection of PCR-amplified HIV-1 DNA. Diagn. Mol. Pathol. *1*:98-102.
30. **Nuovo, G.J., G.A. Gorgone, P. MacConnell, M. Margiotta and P.D. Gorevic.** 1992. In situ localization of PCR-amplified human and viral cDNAs. PCR Methods Appl. *2*:117-123.
31. **Patel, V., A. Shum-Siu, B. Heniford, T. Wieman and F. Hendler.** 1994. Detection of epidermal growth factor receptor mRNA in tissue sections from biopsy specimens using in situ polymerase chain reaction. Am. J. Pathol. *144*:7-14.
32. **Patterson, B., M. Till, P. Otto, C. Goolsby, M. Furtado, L. McBride and S. Wolinsky.** 1993. Detection of HIV-1 DNA and messenger RNA in individual cells by PCR-driven in situ hybridization and flow cytometry. Science *260*:976-979.
33. **Ray, R., P. Komminoth, M. Machado and H. Wolfe.** 1991. Combined polymerase chain reaction and in-situ hybridization for the detection of single copy genes and viral genomic sequences in intact cells. Mod. Pathol. *4*:124A.
34. **Sällström, J., I. Zehbe, M. Alemi and E. Wilander.** 1993. Pitfalls of in situ polymerase chain reaction (PCR) using direct incorporation of labelled nucleotides. Anticancer Res. *13*:1153.
35. **Spann, W., K. Pachmann, H. Zabnienska, A. Pielmeier and B. Emmerich.** 1991. In situ amplification of single copy gene segments in individual cells by the polymerase chain reaction. Infection *19*:242-244.
36. **Staskus, K., L. Couch, P. Bitterman, E. Retzel, M. Zupancic, J. List and A. Haase.** 1991. In situ amplification of visna virus DNA in tissue sections reveals a reservoir of latently infected cells. Microb. Pathog. *11*:67-76.

37. **Tsongalis, G.J., A.H. McPhail, R.R.D. Lodge, J.F. Chapman and L.M. Silverman.** 1994. Localized in situ amplification (LISA): A novel approach to in situ PCR. Clin. Chem. *40*:381-384.
38. **Wijsman, J., R. Jonker, R. Keijzer, C. van de Velde, C. Cornelisse and J. van Dierendonck.** 1993. A new method to detect apoptosis in paraffin sections: In situ end-labeling of fragmented DNA. J. Histochem. Cytochem. *41*:7-12.
39. **Yin, J., M. Kaplitt and D. Pfaff.** 1994. *In situ* PCR and *in vivo* detection of foreign gene expression in rat brain. Cell Vision *1*:58-59.
40. **Zehbe, I., G. Hacker, E. Rylander, J. Sallstrom and E. Wilander.** 1992. Detection of single HPV copies in SiHa cells by in situ polymerase chain reaction (in situ PCR combined with immunoperoxidase and immunogold-silver staining [IGSS] techniques). Anticancer Res. *12*:2165-2168.

Address correspondence to Aidan Long, Department of Medicine, Allergy Division, New England Medical Center Hospitals and Tufts University, 750 Washington Street, Boston, MA 02111, USA.

Applications of *In Situ* PCR Methods in Molecular Biology

O. Bagasra, T. Seshamma, J. Hansen[1], L. Bobroski, P. Saikumari and R.J. Pomerantz

Department of Medicine, Thomas Jefferson University, Philadelphia, PA, and [1]MJ Research, Watertown MA, USA

SUMMARY

The ability to identify individual cells expressing or carrying specific genes of interest in a tissue section under the microscope provides a great advantage in determining various aspects of normal versus pathological conditions. In situ *PCR (ISPCR) can be used in detecting viral infection. In the case of HIV-1 or other viral infection, one can determine the extent of the infection and effects of treatment. Similarly, one could potentially utilize this method in gene therapy for verification of the integration and expression of a desired gene* in vivo. In situ *PCR may also be used to determine tumor burden, before and after chemotherapy, where specific genetic aberrants are associated with certain types of malignancy or to determine the preneoplastic lesions by examining p53 mutations associated with certain types of tumors. In the area of diagnostic pathology, one could find the origin of metastatic tumors by performing reverse transcription* in situ *PCR (RT-PCR). There are about 200 publications describing various forms of* in situ *gene amplifications used to identify various infectious agents, tumor marker genes, cytokines and growth factors and their receptors, as well as in gene therapy. Our laboratories have used ISPCR techniques since 1988, and we have developed a simple, sensitive ISPCR that has proven reproducible in multiple double-blinded studies. One can use this method for amplification of both DNA and RNA gene sequences. By use of multiple-labeled probes, one can perform immunohistochemistry together with RNA and DNA amplification at a single-cell level. To date, we have successfully amplified and detected HIV-1, SIV, HPV, HBV, CMV, EBV, HHV-6, nitrous oxide synthase gene sequences associated with multiple sclerosis by DNA and/or RNA (RT-PCR) in various tissues, including peripheral blood mononuclear cells, lymph nodes, spleen, brain, skin, breast, lungs, cytological specimens, tumors, cultured cells and numerous other formalin-fixed, paraffin-embedded tissues. In this chapter, we provide a detailed account of the ISPCR procedure that can be used for routine research investigation. In addition, we also present detailed procedures for special applications of ISPCR, such as its use in cytogenetics to localize a single gene in the chromosomal bands, its use in dual and triple labeling of cells, where more than one signal can be detected at the single cell level and its combination with electron microscopy, immunohistochemistry and in other special situations.*

INTRODUCTION

Since the publication of the first report regarding the *in situ* amplification of HIV-1 *gag* gene in an HIV-1-infected cell line in 1990 (1), there has been an explosion of research in the area of *in situ* polymerase chain reaction (ISPCR). There are about 200 publications describing various forms of *in situ* gene amplifications (selected bibliography 1–105), identifying various infectious agents (1–17,20,23,32,44,47–78,81–88,96), tumor marker genes (23,78,79,92), cytokines, growth factors and their receptors (37,39,40,65,77) and other genetic elements of interest (18,23,84,94), in peer-reviewed journals. The PCR method for amplification of defined gene sequences has proved a valuable tool, not only for basic researchers but also for clinical scientists (2,4,5,20,23,79–82). Using even a minute amount of DNA or RNA and choosing a thermostable enzyme from a large variety of sources, one can enlarge the amount of the gene of interest, which can be analyzed and/or sequenced. Thus, genes or portions of gene sequences present only in a small sample of cells or small fraction of mixed cellular populations can be examined. However, one of the major drawbacks of the standard PCR technique is that the procedure does not allow the association of amplified signals of the specific gene segment with the histological cell type(s). For example, it would be advantageous to determine what types of cells in the peripheral blood circulation carry HIV-1 provirus at various stages of HIV-1 infection (2–6,62,78) and what percentage of HIV-1-infected cells actually are expressing viral RNA (2–16,35,42). Similar approaches have been used to detect the presence of other gene sequences in tissue materials and pathological specimens (1–105).

The ability to identify individual cells expressing or carrying specific genes of interest in a tissue section under the microscope provides a great advantage in determining various aspects of normal—as opposed to pathological—conditions. For example, this technique could be used in determination of tumor burden before and after chemotherapy in lymphomas or leukemias where specific aberrant gene translocations are associated with certain types of malignancy (23,44,79,80). In the case of HIV-1 infection or other viral infections, one can determine the effects of therapy or putative anti-viral vaccination by evaluating the number of cells still infected with the viral agent post-vaccination. Similarly, one can potentially determine the pre-neoplastic lesions by examining p53 mutations associated with certain tumors or oncogenes or other aberrant gene sequences that are known to be associated with certain types of tumors (2,23,44,79,80). In the area of diagnostic pathology, determination of origin of metastatic tumors is a perplexing problem. By utilizing the proper primers for genes that are expressed by certain histological cell types, one can potentially determine the origin of metastatic tumors by performing reverse transcriptase-initiated *in situ* PCR (80).

Our laboratories have used *in situ* PCR (ISPCR) techniques since 1988 and we have developed a simple, sensitive ISPCR that has proved reproducible in multiple double-blinded studies (1–16,40,52,53,79–82,86,96). One can use this method for amplification of both DNA or RNA gene sequences. By use of multiple labeled probes, one can detect various signals in a single cell. In addition, under special circumstances, one can perform immunohistochemistry and RNA and DNA amplification at a single cell level (the so-called “triple labeling”).

To date, we have successfully amplified and detected human immunodeficiency virus (HIV-1), simian immunodeficiency virus (SIV), human papillomavirus (HPV), hepatitis B virus (HBV), cytomegalovirus (CMV), Epstein barr virus (EBV), human herpes virus-6 (HHV-6), herpes simplex virus (HSV), lymphogranuloma venereum (LGV), p53 and its mutations, mRNA for surfactant Protein A, estrogen receptors, inducible nitric oxide synthatase (iNOS)-gene sequences associated with multiple sclerosis, by DNA and/or RNA (RT ISPCR), in various tissues, including: peripheral blood mononuclear cells

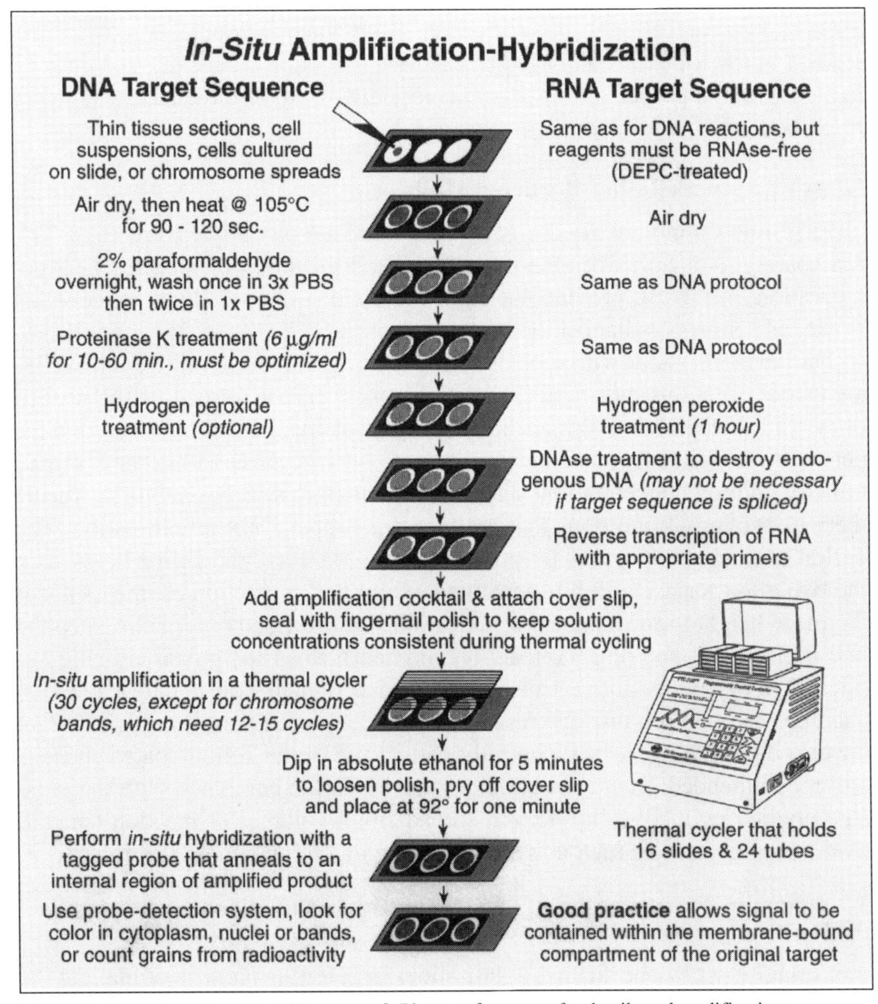

Figure 1. Overview of *in situ* PCR protocol. Please refer to text for details and modifications.

(PBMCs), lymph nodes, spleen, brain, skin, breast, lungs, cytological specimens, tumors, cultured cells and numerous other formalin-fixed, paraffin-embedded tissues. In this chapter, we have provided a detailed account of the ISPCR procedure that can be used for routine research investigations. In addition, we have also provided a detailed procedure for special applications of ISPCR—for example, its use in cytogenetics to localize a single gene in the chromosomal bands; its use in dual and triple labeling of cells, where more than one signal can be detected at the single cell level; and its use in combination with electron microscopy (EM), immunohistochemistry and in other special situations (15).

PREPARATION OF GLASS SLIDES

Before one can perform *in situ* reactions by this protocol, the proper type of glass slide with Teflon® appliqué must be obtained (various sources are described in the Materials and Methods section). Then, the glass surface must be treated with the proper sort of silicon compound. Both of these factors are very important, and the following are reasons why.

Glass Slides with Teflon-Bordered Wells

First, one should always use *glass* slides that are partially covered by a Teflon coating. Not only does the glass withstand the stress of repeated heat denaturation, but it also presents the right chemical surface—silicon oxide—that is needed for proper silanization.

Furthermore, slides with special Teflon coatings that form individual "wells" are useful because vapor-tight reaction chambers can be formed on the surface of the slides when coverslips are adhered with coatings of nail polish around the periphery. These reaction chambers are necessary because within them proper tonicity and ion concentrations can be maintained in aqueous solutions during thermal cycling—conditions that are vital for proper DNA amplification. The Teflon coating serves a dual purpose in this regard. First, the Teflon helps keep the two glass surfaces slightly separated, allowing for reaction chambers about 20 μm in height to form in between. Secondly, the Teflon border helps keep the nail polish from entering the reaction chambers when the polish is being applied. This is important, because any leakage of nail polish into a reaction chamber can compromise the results in that chamber. Even if one is using an advanced thermal cycler with humidification, use of the Teflon-coated slides is still recommended. The hydrophobicity of the Teflon combined with the pressure applied by a coverslip help to spread small volumes of reaction cocktail over the entire sample region, without forcing much fluid out the periphery.

3-Aminopropyltriethoxysilane (AES) Silanization: Putting on the Positive Charge

Normally, when one silanizes glass slides, it is for the purpose of making the slide surface hydrophobic. However, the AES silane that is specified in this protocol has a very different surface effect. This silicon compound imparts a strong, positive and persistent electrical charge to the slide by forming an aminopropyl derivative of glass at the surface of the slide. The resulting positive charge—and electrostatic attraction—causes the cells or tissues to adhere with great tenacity throughout the amplification-hybridization procedure. Experiments have been conducted with slides coated with alternative adhesives, including white glue, albumen, chrome-gelatin and poly-L-lysine. However, we have found that slides treated with AES silane have superior tissue adhesion and lower background with all tissue types tested, including mixed cell suspensions, paraffin sections and frozen sections. We hypothesize that this adhesion occurs because the negatively charged molecules on membrane surfaces ionically bond to the positively charged surface of the glass when the cells or tissues are initially placed on the slide. However, the limited positive charge is probably canceled rather quickly, so that later in the procedure when one adds solutions that contain additional negatively charged molecules, the new molecules float freely without becoming bound to the slide surface.

In order to prepare glass slides properly, follow this procedure:

1. Prepare the following 2% AES solution just prior to use:

3-AES (A-3648; Sigma Chemical, St. Louis, MO, USA)	5 mL
Acetone	250 mL

2. Put solution into a Coplin jar or glass staining dish and dip glass slides in 2% AES for 60 s (see Materials and Methods section for sources of both Coplin jars and the proper glass slides).
3. Dip slides five times into a different vessel filled with 1000 mL of distilled water.
4. Repeat step #3 three times, changing the water each time.
5. Air-dry in laminar-flow hood from a few hours to overnight, then store slides in sealed container at room temperature. Try to use slides within 15 days of silanization; 250 mL of AES solution is sufficient to treat 200 glass slides.

PREPARATION OF TISSUE

Cell Suspensions

To use peripheral blood leukocytes, first isolate cells on a Ficoll®-Hypaque density gradient. Tissue-culture cells or other single-cell suspensions can also be used. Prepare all cell suspensions with the following procedure:

1. Wash cells with 1× phosphate-buffered saline (PBS) twice.
2. Resuspend cells in PBS at 2×10^6 cells/mL.
3. Add 10 μL of cell suspension to each well of the slide using a P20 micropipet. Spread suspension across well surface.
4. Air-dry slide in a laminar-flow hood.

Paraffin-Fixed Tissue

Routinely fixed paraffin tissue sections can be amplified quite successfully. This permits the evaluation of individual cells in the tissue for the presence of a specific RNA or DNA sequence. For this purpose, tissue sections are placed on specially designed slides that have *single* wells—which are described further in the Materials and Methods section. In our laboratory, we routinely use placental tissues, central nervous system (CNS) tissues, cardiac tissues, etc., which are sliced to a 3–5 µm thickness. Other laboratories prefer to use sections up to 10 µm thick, but in our experience, amplification is often less successful with the thicker sections, and multiple cell layers can often lead to difficult interpretation due to superposition of cells. However, if one is using tissues that contain particularly large cells—such as ovarian follicles—then thicker sections may be appropriate.

1. Place tissue section upon the glass surface of the slide.
2. Incubate the slides in an oven at 60°–80°C (depending on type of paraffin used to embed the tissue) for 1 h to melt the paraffin.
3. Dip the slides in EM grade xylene solution for 5 min, then in EM grade 100% ethanol for 5 min (EM grade reagents are benzene-free). Repeat these washes two or three times in order to remove the paraffin completely.
4. Dry the slides in an oven at 80°C for 1 h.

Discussion on Frozen Sections

It is possible to use frozen sections for *in situ* amplification; however, the morphology of the tissue following the amplification process is generally not preserved as well as with paraffin sections. It seems that the cryogenic freezing of the tissue, combined with the lack of paraffin substrate during slicing, compromises the integrity of the tissue. Usually, thicker slices must be made, and the tissue "chatters" in the microtome. As any clinical pathologist will relate, definitive diagnoses are made from *paraffin* sections, and this rule-of-thumb seems to extend to the amplification procedure as well.

It is very important to use tissues that are frozen in liquid nitrogen or placed on dry-ice immediately after they are harvested before autolysis begins to take place. If tissues are frozen slowly by placing them in -70°C freezer, eventually ice crystals will form inside the tissues, creating a gap that will distort the morphology.

For frozen sections, one should use as thin a slice as possible (down to 3–4 µm), apply to slide, dehydrate for 10 min in 100% methanol (exception to methanol is when surface antigens are lipoprotein and will denature in methanol, then use 2% paraformaldehyde or other reagent) and air-dry in a laminar-flow hood. Then, proceed with heat treatment described below.

IN SITU AMPLIFICATION: DNA AND RNA TARGETS

Basic Preparation, All Protocols

For all sample types, the following steps comprise the basic preparatory work that must be done before any amplification-hybridization procedure.

Heat Treatment

Place the slides with adhered tissue on a heat-block at 105°C for 5–30 s, to stabilize the cells or tissue on the glass surface of the slide.

This step is absolutely critical, and one may need to experiment with different periods in order to optimize the heat treatment for specific tissues. Our laboratory routinely uses 90 s for DNA target sequences and 5–10 s for RNA sequences. The shorter incubation is recommended for RNA targets because certain mRNAs may be unstable at high temperatures.

Fixation and Washes

1. Place the slides in a solution of 4% paraformaldehyde in PBS (pH 7.4) for 4 h at room temperature. Use of the recommended Coplin jars or staining dishes facilitates these steps.
2. Wash the slides once with 3× PBS for 10 min, agitating periodically with an up and down motion.
3. Wash the slides with 1× PBS for 10 min agitating periodically with an up and down motion. Repeat once with fresh 1× PBS.
4. At this point, slides with adhered tissue can be stored at -80°C until use. Before storage, dehydrate with 100% ethanol.

If biotinylated probes or peroxidase-based color development are to be used, the samples should further be treated with a 0.3% solution of hydrogen peroxide in PBS, in order to inactivate any endogenous peroxidase activity. Once again, incubate the slides overnight—either at 37°C or at room temperature. Then, wash the slides once with PBS.

If other probes are to be used, proceed directly to the following proteinase K digestion, which is perhaps the most critical step in the protocol.

Proteinase K Treatment (the Most Rate-Limiting Step)

1. Treat samples with 6 µg/mL proteinase K in PBS for 5 to 60 min at room temperature or at 55°C. To make a proper solution, dilute 1.0 mL of proteinase K at 1 mg/mL in 150 mL of 1× PBS.
2. After 5 min, look at the cells under the microscope at 400×. If the majority of the cells of interest exhibit uniform-appearing, small, round "salt-and-pepper" dots on the cytoplasmic membrane, then stop the treatment immediately with Step #3. Otherwise, continue treatment for another 5 min and re-examine.
3. After proper digestion, heat slides on a block at 95°C for 2 min to inactive the proteinase K.

4. Rinse slides in 1× PBS for 10 s.
5. Rinse slides in distilled water for 10 s.
6. Air-dry.

Discussion on Optimizing Digestion

The time and temperature of incubation should be optimized carefully for each cell line or tissue-section type. With too little digestion, the cytoplasmic and nuclear membranes will not be sufficiently permeable to primers and enzyme, resulting in inconsistent amplification. With too much digestion, the membranes will either deteriorate during repeated denaturation or worse, the signals will leak out. In the first case, cells will not contain the signal and high background will result. In the latter case, many cells will show peri-cytoplasmic staining, representing the leaked signals going into the cells not containing the signals. Attention to detail here can often mean the difference between success and failure, and this procedure should be practiced rigorously with extra sections before continuing on to the amplification steps.

In our laboratory, proper digestion parameters vary considerably with tissue type. Typically, lymphocytes will require 5–10 min at 25°C or room temperature, CNS tissue will require 12–18 min at room temperature and paraffin-fixed tissue will require between 15–30 min at room temperature. However, these periods can vary widely and the appearance of the "salt and pepper" dots is the important factor. Unfortunately, the appearance of the salt and pepper dots is less prominent in paraffin sections. The representative examples of salt and pepper dots on cells from cell suspension and from a brain tissue section are illustrated in Figure 2 (A and B).

We have also experimented with the use of other proteinases—such as amylase, trypsin, pronase and pepsin—instead of proteinase K. This has proven successful in many circumstances; for example, we incubate slides for 10 min with pepsinogen at 1 mg/mL at pH 2.0. However, we have found that proteinase K almost always gives better overall results.

The critical importance of these dots should not be underestimated, since an extra 2–3 min of treatment after the appearance of dots will result in leakage of signals.

An alternative to the observation of the "dots" method is to select a constant time and treat slides in varying amounts of proteinase K. For example, treat slides for 15 min in 1–6 µg/mL of proteinase K.

Reverse Transcription (RT) Variation: *In Situ* RNA Amplification

There are two choices to detect an RNA signal. The first and more elegant method is simply to use primer pairs that flank spliced sequences of mRNA; these particular sequences will be found *only* in RNA and will be split into sections in the DNA (see Figure 3). Thus, by using these RNA-specific primers, one can skip the following DNase step and proceed directly to reverse transcription.

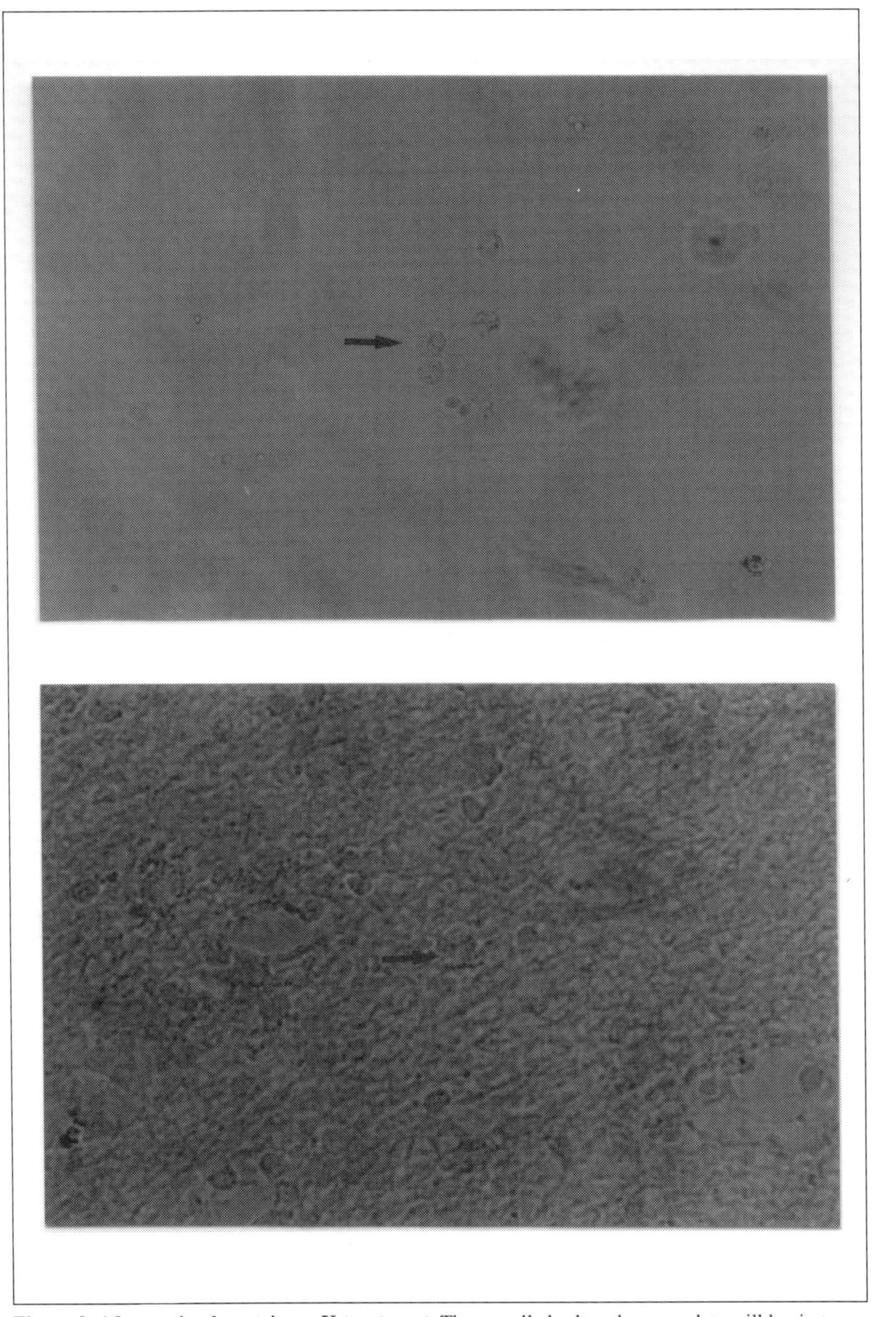

Figure 2. After optimal proteinase K treatment. The so-called salt and pepper dots will begin to appear approximately 5 min after the initiation with proteinase K treatment. The optimal period for the enzymatic treatment is determined when there are around 10–20 "peppery dots" on the cell surface. The best way to visualize the "dots" is to use phase-contrast microscopy. As illustrated above in **A)**, there are numerous cells with the peppery dots (arrows). All the cells in this visual field exhibited the presence of the dots. In **B)** a representative section from a brain section is shown. Here the visualization is difficult (arrows) and requires much practice. *(A color reproduction of this figure appears on p. 139.)*

The other, more brutal, yet often necessary, approach is to treat the cells or tissue with a DNase solution subsequent to the proteinase K digestion. This step destroys all of the endogenous DNA in the cells so that only RNA survives to provide signals for amplification.

Note: All reagents for RT *in situ* amplification should be prepared with RNase-free water (i.e., diethyl pyrocarbonate [DEPC]-treated water). In addition, the silanized glass slides and all glassware should be RNase-free. We ensure this by baking the glassware overnight in an oven before use in the RT-amplification procedure.

DNase Treatment

Prepare a RNase-free, DNase solution:

40 mM Tris-HCl, pH 7.4

6 mM $MgCl_2$

2 mM $CaCl_2$

1 U/mL final volume of DNase (use an RNase-free DNase, such as 10 U/mL RQ1 DNase; Cat. # 776785 from Boehringer Mannheim, Indianapois, IN, USA)

1. Add 10 μL of solution to each well.
2. Incubate the slides overnight at 37° in a humidified chamber. If one is using liver tissue, this incubation should be extended an additional 18–24 h.
3. After incubation, rinse the slides with a similar solution prepared without the DNase enzyme. Heat slides to 92° for 2 min to inactivate enzyme.
4. Wash the slides twice with DEPC-treatment water.

Note: Some cells are particularly rich in ribonuclease; in this circumstance, add the following ribonuclease inhibitor to the DNase solution: 1000 U/mL placental ribonuclease inhibitor (e.g., RNasin®) plus 1 mM dithiothreitol (DTT).

RT Reaction

Next, one wishes to make DNA copies of the targeted RNA sequence so that the signal can be amplified. The following are typical cocktails for the RT reaction:

If using AMVRT or MMLVRT enzyme:

10× reaction buffer (see below)	2.0 μL
10 mM dATP	2.0 μL
10 mM dCTP	2.0 μL
10 mM dGTP	2.0 μL
10 mM dTTP	2.0 μL
40 U/μL RNasin (Promega, Madison, WI, USA)*	0.5 μL
20 μM downstream primer	1.0 μL
AMVRT 20 U/ μL	0.5 μL

DEPC-treated water	8.0 μL
Total Volume:	20.0 μL

10× reaction buffer: 100 mM Tris pH 8.3, 500 mM KCl, 15 mM $MgCl_2$.

If using SuperScript™ II enzyme from Life Technologies (Gaithersburg, MD, USA):

5× reaction buffer (as supplied with enzyme)	4.0 μL
10 mM dATP	2.0 μL
10 mM dCTP	2.0 μL
10 mM dGTP	2.0 μL
10 mM dTTP	2.0 μL
RNasin at 40 U/μL (Promega)*	0.5 μL
20 μM downstream primer	1.0 μL
SuperScript II, 200 U/μL	0.5 μL
0.1 M DTT	1.2 μL
DEPC-treated water	4.8 μL
Total Volume:	20.0 μL

*RNasin inhibits ribosomal RNases—use for optimal yields.

1. Add 10 μL of either cocktail to each well. Carefully place the coverslip on top of the slide.
2. Incubate at 42°C or 37°C for 1 h in a humidified atmosphere.
3. Incubate slides at 92°C for 2 min.
4. Remove coverslip and wash twice with distilled water. Proceed with the amplification procedure, which is the same for both DNA- and RNA-based protocols.

Discussion of Reverse Transcriptase Enzymes

Avian myoblastosis virus reverse transcriptase (AMVRT) and Moloney murine leukemia virus reverse transcriptase (MMLVRT) give comparable results in our laboratory. Other reverse transcriptase enzymes will probably work also. However, it is important to read the manufacturers' descriptions of the reverse transcriptase enzyme and to make certain that the proper buffer is used.

An alternative reverse transcriptase enzyme is available that lacks RNase activity. Called SuperScript II, it is suitable for RT of long mRNAs. It is also suitable for routine reverse transcriptase amplification and in our laboratory it has proven to be more efficient than the two enzymes described above.

Discussion of Primers and Target Sequences

In our laboratory, we simply use antisense downstream primers for our gene of interest, as we already know the sequence of most genes we study. However, one can alternatively use oligo (dT) primers to first convert all mRNA populations into cDNA, and then perform the *in situ* amplification for a specific cDNA. This technique may be useful when one is performing amplification of several different gene transcripts at the same time in a single cell. For example, if one is attempting to detect various cytokine expressions, one can use an oligo (dT) primer to reverse transcribe all of the mRNA copies in a cell or tissue section. Then, one can amplify more than one type of cytokine and detect the various types with different probes that develop into different colors (see section on multiple signals in individual cells).

In all RT reactions, it is advantageous to reverse transcribe only relatively small fragments of mRNA (<300 bp). Larger fragments may not completely reverse transcribe due to the presence of secondary structures. Furthermore, the reverse transcriptase enzymes—AMVRT and MMLVRT, at least—are not very efficient in transcribing large mRNA fragments. However, this size restriction does not apply to amplification reactions that are exclusively DNA, for the polymerase enzyme copies nucleotides better. In *in situ* DNA reactions, we routinely amplify genes up to 800 bp.

The following are several additional points one should keep in mind:

(i) The length for both sense and antisense primers should be 14–22 bp.

(ii) At the 3′ ends, primers should contain a GC-type base pairs (e.g., GG, CC, GC or CG) to facilitate complementary strand formation.

(iii) The preferred GC content of the primers is from 45%–55%.

(iv) Try to design primers so they do not form intra- or interstrand base pairs. Furthermore, the 3′ ends should not be complementary to each other or they will form primer dimers.

One can design a reverse transcriptase primer so that it does not contain secondary structures.

Discussion of Annealing Temperatures

Annealing temperatures for reverse transcription and for DNA amplification can be chosen according to the following formula:

$$T_m \text{ of the primers} = 81.5°C + 16.6 (\text{Log } M) + 0.41 (G + C\%) - 500/n$$

where n = length of primers and M = molarity of the salt in the buffer, usually 0.047 M for DNA reactions and 0.070 M for RT reactions (see below). If using AMVRT, the value will be lower according to the following formula:

$$T_m \text{ of the primers} = 62.3°C + 0.41 (G + C\%) - 500/n$$

Usually, primer annealing is optimal at 2°C above its T_m. However, this formula provides only an approximate temperature for annealing, since base-stacking, near-neighbor effect and buffering capacity may play a significant role for a particular primer.

Optimization of the annealing temperature should be carried out first with solution-based reactions. It is important to know the optimal temperature before attempting to conduct *in situ* amplification, as *in situ* reactions are simply not as robust as solution-based ones. We hypothesize that this is due to the fact that primers do not have easy access to DNA templates inside cells and tissues, as numerous membranes, folds and other small structures can keep primers from binding homologous sites as readily as they do in solution-based reactions.

There are two additional ways to determine the real annealing temperatures:

1. To utilize a recently developed thermocycler designed for determination of actual annealing temperature called Robocycler™ Gradient 40 temperature Cycler from Stratagene (La Jolla, CA, USA).
2. Another method is to utilize so-called "Touchdown" PCR (22).

The logic of determining the correct annealing temperature for IS-PCR is as follows:

During amplification, spurious products often appear in addition to those desired. Therefore, even if the cells do not contain DNA homologous to the primer sequences, many artifactual bands may appear. Many protocols have appeared in the literature to overcome false priming, including hot-start (63,67,75), use of dimethyl sulfoxide (DMSO), formamide and anti-*Taq* antibodies (75; i.e., TaqStart™; Clontech, Palo Alto, CA, USA). An important thing to remember is that false priming will occur if melting temperature (T_m) between primer and template is not accurate. Therefore, temperature above the T_m will yield no products and temperature too far below the T_m often will give unwanted products due to false priming. Therefore determination of optimal annealing temperature is extremely important.

Recently, MJ Research (Watertown, MA, USA) devised a thermocycler that has the capcity to perform both *in situ* gene amplification in slides and in solution (tubes) simultaneously, in the same block. That kind of thermocycler can be very useful in determining the optimal amplification of a gene of interest.

Amplification Protocol, All Types

Prepare an amplification cocktail containing the following: 1.25 μM of each primer, 200 μM (each) deoxyribonucleoside triphosphate (dNTP), 10 mM Tris-HCl (pH 8.3), 50 mM KCl, 2.5 mM $MgCl_2$, 0.001% gelatin and 0.1 U/μL *Taq* DNA polymerase. The following is a convenient recipe that we use in our laboratory:

25 μM forward primer (i.e., SK 38 for HIV-1)	5.0 μL
25 μM reverse primer (i.e., SK 39 for HIV-1)	5.0 μL
10 mM each dNTP	2.5 μL
1.0 M Tris-HCl pH 8.3	1.0 μL
1.0 M KCl	5.0 μL
100 mM $MgCl_2$	2.5 μL
0.01% gelatin	10.0 μL
Taq pol (AmpliTaq® 5 U/μL; Perkin-Elmer,	

Norwalk, CT, USA)* 2.0 μL

H_2O 67.0 μL

Total Volume: 100.0 μL

Note: Other thermostable polymerase enzymes have also been used quite successfully.

1. Layer 10 μL of amplification solution onto each well with a P20 micropipet so that the whole surface of the well is covered with the solution. Be careful—do not touch the surface of the slide with the tip of the pipet.
2. Add a glass coverslip (22 × 60 mm) and carefully seal the edge of the coverslip to the slide with clear nail polish or varnish. If using tissue sections, use a second slide instead of a cover slip (see discussions below on attaching coverslips and hot start).
3. Place slides on a heat-block at 92°C for 90 s.
4. Place slides in a thermocycling instrument.
5. Run 30 cycles of the following amplification protocol:

94°C	30 s
45°C	1 min
72°C	1 min

These times/temperatures will likely require optimization for the specific thermocycler being used. Furthermore, the annealing temperature should be optimized, as described earlier. These particular incubation parameters work well with SK38 and SK39 primers for the HIV-1 *gag* sequence, when amplified in an MJ Research PTC-100™-60 or PTC-100-16MS thermal cycler.

6. After the thermal cycling is complete, dip slides in 100% EtOH for at least 5 min, in order to dissolve the nail polish. Pry off the coverslip using a razor or other fine blade—the coverslip generally pops off quite easily. Scratch off any remaining nail polish on the outer edges of the slide so that fresh coverslips will lay evenly in the subsequent hybridization/detection steps.
7. Place slides on a heat-block at 92°C for 1 min—this treatment helps immobilize the intracellular signals.
8. Wash slides with 2× SSC (see Materials and Methods section) at room temperature for 5 min.

The amplification protocol is now complete and one can proceed to the labeling/hybridization procedures.

Discussion on Attaching Coverslip/Top Slide

Be certain to carefully paint the polish around the entire periphery of the coverslip or the edges of the dual slide, as the polish must completely seal the coverslip-slide assembly in order to form a small reaction "chamber" that can contain the water vapor during thermal cycling. For effective sealing, do not use colored polish or any other nail polish which is especially "runny" —our laboratory prefers to use Wet & Wild Clear nail polish. Proper sealing is very important to keep reaction concentrations consistent throughout the thermal cycling procedure. Concentrations of reagents are critical for proper amplification. However, be certain to apply the nail polish very carefully so that none of the polish gets into the actual chamber where the cells or tissues reside. If any nail polish does enter the chamber, discard that slide because the results will be questionable. Please bear in mind that the painting of nail polish is truly a *learned* skill; therefore, it is strongly recommended that researchers practice this procedure several times with mock slides before attempting an experiment.

In the case of tissue sections, it is best to use another identical blank slide for the cover instead of a coverslip. Apply the amplification cocktail to the appropriate well of the blank slide, place an inverted tissue-containing slide on top of the blank slide and seal the edges as described. Invert the slide once again so that the tissue-containing slide is on the bottom. This technique can be modified to accommodate a hot start (see discussion below).

Discussion of Hot Start Technique

There is much debate as to whether a hot start helps to improve the specificity and sensitivity of amplification reactions. In our laboratory, we find the hot start adds no advantage in this regard; rather, it adds only technical difficulty to the practice of the *in situ* technique. However, a variation on the "hot start" has been reported recently (42,75). In this procedure, one simply uses anti-*Taq* antibody in the PCR cocktail (containing *Taq*), which keeps the *Taq* enzyme in the cocktail "blocked" until the first cycle of 92°C when anti-*Taq* antibody gets denatured and restores the full *Taq* activity. This modification essentially serves the same function as the "hot start" procedure but without its difficulties (75).

Discussion of Thermal Cyclers

Various technologies of thermocycler will work in this application; however, some instruments work much better than others. In our laboratory, we use two types: a standard, block-type thermocycler that normally holds sixty 0.5-mL tubes but that can be adapted with aluminum foil, paper towels and a weight to hold 4–6 slides. We also use dedicated thermocyclers that are specifically designed to hold 12 or 16 slides. We understand that other labs have used stirred-air, oven-type thermocyclers quite successfully; however, we have also heard that there are sometimes problems with the cracking of glass slides during cycling. Thermocyclers dedicated to glass slides are now available from several vendors, including Barnstead (Iowa), Coy Corporation (Grass Lake, MN, USA), Hybaid (Teddington, Middlesex, UK), and MJ Research. Our laboratory has used an MJ Research PTC-100-12MS and a PTC-100-16MS quite successfully. Recently, this company has combined the slide and tubes into a single block, allowing the simultanous confirmation of *in situ* amplification in a tube. There are newer designs of thermal cyclers which incorporate humidification chambers; however, we do not yet have sufficient experience with this technology to verify whether they can eliminate the need for sealing the slides with nail polish during thermal cycling. Nonetheless, the humidified instruments are especially useful in the RT and hybridization steps, where otherwise a humidified incubator is needed.

We suggest that you follow the manufacturer's instructions on the use of your own thermocycler, bearing in mind the following points:

(i) Glass does not easily make good thermal contact with the surface on which it rests; therefore, a weight to press down the slides and/or a thin layer of mineral oil to fill in the micro air pockets will help thermal conduction. If using mineral oil, make certain that the oil is well smeared over the glass surface so that the slide is not merely floating on air bubbles beneath it.

(ii) The top surfaces of slides lose heat quite rapidly through radiation and convection; therefore, use a thermocycler that envelops the slide in an enclosed chamber (as in some dedicated instruments), or insulate the tops of the slides in some manner. Insulation is particularly critical when using a weight on top of the slides, for the weight can serve as an unwanted heat sink if it is in direct contact with the slides.

(iii) Good thermal uniformity is imperative for good results. Poor uniformity or irregular thermal change can result in cracked slides, uneven amplification or completely failed reactions. If adapting a thermocycler that normally holds plastic tubes, use a layer of aluminum foil to spread out the heat.

One-Step RT Amplification

If one uses reverse transcriptase enzymes to manufacture cDNA that is subsequently amplified by *Taq* or *Vent* polymerase, one must use two different buffer systems—one solution for the RT and another for the DNA amplification. However, one can use a single, recombinant enzyme r*Tth*, which can do both jobs at once. The typical cocktail for this single-step reaction is the following:

100 μM forward primer	0.5 μL
100 μM reverse primer	0.5 μL
3 mM nucleotide mixture (dNTP)	6.0 μL
10 mM $MnCl_2$	2.0 μL
25 mM $MgCl_2$	10.0 μL
10× transcription buffer (below)	2.0 μL
10× chelating buffer (below)	8.0 μL
1.7 mg/mL BSA	10.0 μL
2.5 U/mL r*Tth* enzyme	2.0 μL
DEPC-treated water	59.0 μL
Total Volume:	100.0 μL

10× Transcription buffer: 100 mM Tris-HCl pH 8.3, 900 mM KCl.

10× Chelating buffer: 100 mM Tris-HCl pH 8.3, 1 M KCl, 7.5 mM EGTA, 0.5% Tween™-20, 50% (vol/vol) glycerol.

This reaction requires a slightly variant thermal cycling profile. Our laboratory uses the following amplification protocol:

70°C	15 min
92°C	3 min
70°C	15 min
92°C	3 min
70°C	15 min

Then, 29 cycles of the following profile:

93°C	1 min
53°C	1 min
72°C	1 min

Direct Incorporation of Nonradioactive Labeled Nucleotides

Several nonradioactive labeled nucleotides are available from various sources (i.e., dCTP-biotin, digoxin II-dUTP, etc.). These nucleotides can be used to directly label amplification products, and then the proper secondary agents and chromogens can be used to detect the directly labeled *in situ* amplification products (see below). However, in our opinion—as well as in the opinion of several other laboratory groups—the greatest specificity is *only* achieved by conducting amplification followed by subsequent *in situ* hybridization (1–16,20,42,44,45,48,57,97,103). In the direct labeling protocols, nonspecific incorporation can be significant, and even if this incorporation is minor, it still leads to false-positive signals similar to non-specific bands in gel electrophoresis following solution-based DNA or RT amplification. We strongly discourage the direct incorporation of labeled nucleotides as part of an *in situ* amplification protocol.

The only exception to this recommendation is when one is screening a large number of primer pairs for optimization of a specific assay—then, direct incorporation may be useful. To perform such screenings, add to the amplification cocktail detailed earlier the following: 4.3 μM labeled nucleotide—either 14-biotin dCTP, 14-biotin dATP or 11-digoxigenin dUTP—along with cold nucleotide to achieve a 0.14 mM final concentration. Also, if one has worked out the perfect annealing system, either using a Robocycler or equivalent system, then one can use direct incorporation without fear of nonspecific labeling, which we have discussed elsewhere in detail (8).

LABELING OLIGONUCLEOTIDE PROBE

^{33}P Labeling of Probe

Many laboratories wish to use a radioactively labeled probe, and the following is a typical procedure for the ^{33}P labeling of a probe through a kinase reaction:

2 µM probe (SK 19)	1.0 µL
10 kinase buffer	2.0 µL
ATP γ ^{33}P (10 µCi/µL; Amersham, Arlington Heights, IL, USA)	1.0 µL
H_2O	15.0 µL
Polynucleotide kinase (10 U/µL)	1.0 µL
Total Volume:	20.0 µL

1. Incubate at 37°C for 30 min.
2. Apply sample to 0.8 mL Sephadex® G-50 column (e.g., QuickSpin™ from Boehringer Mannheim)
3. Elute with Tris EDTA buffer.

Fraction #1	200 µL
Fraction #2	100 µL
Fraction #3	100 µL
Fraction #4	100 µL
Fraction #5	100 µL
Fraction #6	100 µL

4. Count the radioactivity in 1.0 µL of each fraction. The labeled probe should be contained in Fraction 2 through Fraction 4.

Discussion on Use of ^{35}S-Labeled Nucleotides

Several manufacturers of thermal cyclers, as well as suppliers of labeled nucleotides, have reported contamination problems associated with a chemical breakdown product from ^{35}S-labeled nucleotides. It is hypothesized that when these particular compounds are subject to high temperatures, a gaseous, radioactive breakdown product forms—possibly H_2S—which leaks from the sample vessel. This can lead to contamination of the incubating device and, possibly, the air in the laboratory. Therefore, we recommend the use of nucleotides labeled with the newer phosphorus isotope, ^{33}P, rather than ones labeled with ^{35}S.

Nonradioactive Labeling-Tailing with DIG-11-dUTP

Many laboratories—including our own—prefer to use nonradioactive probes, for this results in less hazardous waste, fewer bureaucratic procedures and lower costs (especially when one considers the perishable nature of radioactive isotopes and probes). The following is a typical labeling procedure

20 µM Probe	10.0 µL
5× Tailing buffer (see below)	10.0 µL
25 mM $CaCl_2$	20.0 µL
2.5 mM dATP (Tris buffer, pH 7.5)	3.5 µL
1 mM DIG-11-dUTP	1.0 µL
H_2O	4.5 µL

Terminal Transferase (25 U/μL) 1.0 μL
Total Volume:50.0 μL

Tailing buffer (5×): 1 mM potassium cacodylate, 125 mM Tris-HCl pH 6.6 and 1.25 mg/mL bovine serum albumin (BSA).

Incubate reaction mixture at 37°C for 15 min, then purify the labeled probe as for described above for radiolabeled probes, using a chromogen indicator instead of radioactive detectors.

We prefer to use biotinylated fluorescein-isothiocyanale (FITC) or alkaline phosphotase probes, which we purchase already conjugated. These probes are usually not tail-end conjugated; rather, they had been conjugated during the oligo-synthesis procedure. Thus, they may contain multiple molecules of biotin instead of one at the tail end, which makes the probes more sensitive.

Probes are good for up to one year at -70°C.

SPECIAL APPLICATION OF *IN SITU* AMPLIFICATION

EM

Our group at Thomas Jefferson and another in Uppsala University in Sweden have been developing techniques to observe the results of *in situ* amplification under EM. The procedure both of us have used is a simple modification of the immunogold EM technique.

First, the *in situ* amplification is carried out in solution rather than with the cells or tissue adhered to a glass slide. The cells or tissue are fixed in 4% paraformaldehyde for 4 h, washed in PBS buffer and then treated with proteinase K. As before, the cells are observed under an optical microscope for the development of "bubbles" or "salt and pepper" dots, as described in the discussion on optimizing digestion. Amplification of DNA or RNA is then carried out, as described earlier.

In order to check the validity of the results, a small portion of the cells is withdrawn and placed on a slide and *in situ* hybridization performed with the specific biotinylated probe and an unrelated biotinylated probe (as a negative control). The color is developed with streptavidin-peroxidase or streptavidin-alkaline phosphatase. If the color develops with the specific probe and the nonspecific control is negative, then immunolabeling can be performed with the remaining cells.

For this purpose, cells are hybridized with biotinylated probe(s) and then labeled with streptavidin-immunogold conjugates. After 1 h incubation at 37°C, unbound conjugates are washed extensively and cells pelletized by centrifugation. The cell pellet now can be incubated in freshly prepared 2.5% gluteraldehyde solution and processed for EM work.

EM grids can also be used as the substrate for the cells or tissues. Place the grid between two slides and perform the *in situ* amplification steps as described above. Then, perform the labeling with the biotin-streptavidin-immunogold steps.

In our hands, we find the harsh treatment inflicted by repeated denaturation tends to destroy the internal organelles of cells. However, clear signals can be detected in the perinuclear areas of the cells.

In Situ Amplification and Immunohistochemistry

Immunohistochemistry and *in situ* amplification can be performed simultaneously in a single cell. For this purpose, we first fix cells or frozen sections of tissue, which are already placed on slides, with 100% methanol for 10 min. Then, slides are washed in PBS. After that, labeling of surface antigen(s) can be carried out by standard immunohistochemistry methods (i.e., primary

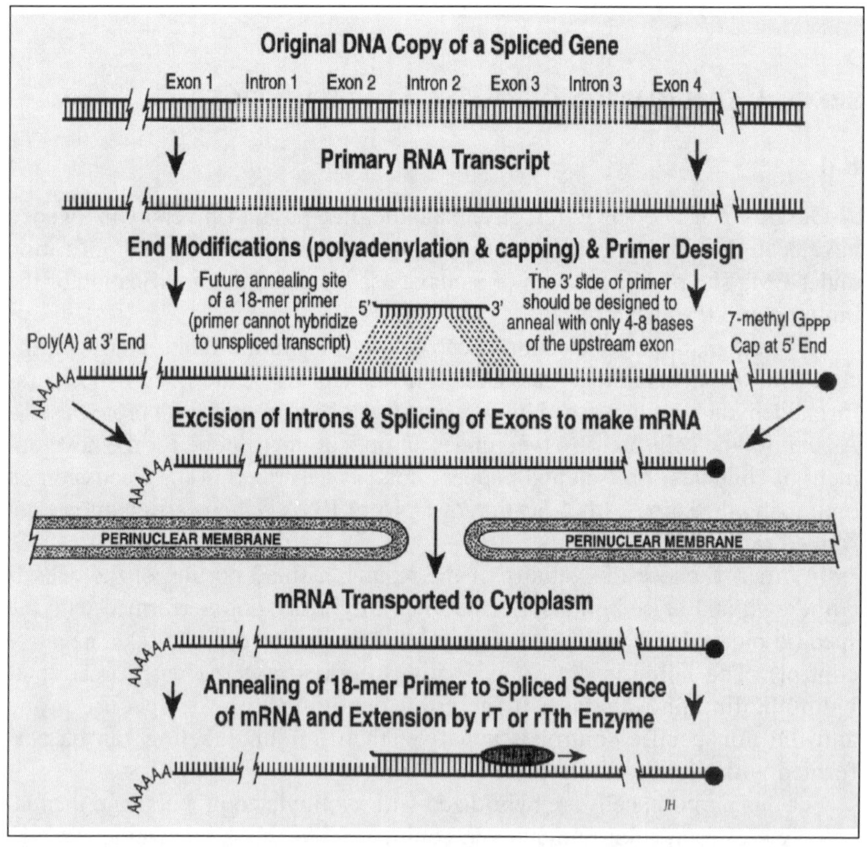

Figure 3. Most eukaryotic genes are split into segments, as there are numerous "introns" in the DNA that are excised during the synthesis of mRNA in the nucleus. This characteristic can be exploited in the design of primers in order to amplify mRNA signals without interference from the DNA. Simply design primers so that their sequences flank spliced regions where two exons are fused, such that the homologous annealing sites exist only in mRNA, not in DNA. This allows the elimination of a DNase treatment during slide preparation, as well as the simultaneous amplification of both RNA and DNA signals.

unconjugated antibody is incubated for 1 h at 37°C, and washed, and then cells or tissue sections are fixed in 4% paraformaldehyde for 2 h). In various pathology laboratories, many specific surface antigens have been tabulated that can withstand 10% formalin and other routine histopathology procedures and will still bind specific monoclonal antibodies. If one is using any of these immunohistochemistry panels, then one can use routinely prepared paraffin sections for the detection of cellular antigens. Then, the tissue is prepared for *in situ* amplification, as described earlier.

Following the development of the color of the amplified product in the post-hybridization step, one can view the cells or tissue under UV/visible light in an alternating manner to detect two signals in a single cell. Secondary antibody (FITC-labeled) after color development of amplified product is used to visualize the antibody binding site.

Multiple Signals, Multiple Labels in Individual Cells

DNA, mRNA and protein can all be detected simultaneously in individual cells. As described in the above section, one can label proteins by FITC-labeled antibodies. Then, one can perform both RNA and DNA *in situ* amplification in the cells. If one is using primers for spliced mRNA and if these primers are not going to bind any sequences in DNA, then both DNA and RT amplification can be carried out simultaneously. Of course, one still needs to perform an RT step, but this time without pre-DNase treatment. In Figure 3, we have illustrated how we design such primers. Subsequently, products can be labeled with different kinds of probes, resulting in different colors of signal. For example, proteins can have a FITC-labeled probe, mRNA can show a rhodamine signal (rhodamine-conjugated probe; 20 colors are available) and DNA can be detected with a biotin-labeled probe. Each will show a different signal within an individual cell.

In Situ Amplification on Chromosomal Bands

Our laboratory has successfully amplified an HIV-1 sequence on chromosomal bands prepared from SUP-T1 infected cell lines. We have modified the chromosomal banding procedure so that it can be used for *in situ* gene amplification. The basic principles are the following: (1) carry out chromosomal banding on the specially designed slides; (2) cover the naked chromosomes with cell ghost membranes so that the signal will remain localized; (3) use fewer cycles of amplification (10–12) so the signal will not leak out; and (4) eliminate the heat fixation step. Currently, we are using this technique for PBMCs from HIV-1 infected individuals. The following is a detailed procedure.

PRECAUTIONS

1. Sterile technique must be practiced at all times in this procedure. This is necessary in handling cell cultures, both to protect the investigator and to avoid

introducing microbial contamination of the cell culture system. Such contamination is often the cause of test failure.

2. The human peripheral blood used in this procedure may be infectious or hazardous to the investigator. Proper handling and decontamination and disposal of waste material must be emphasized.

a) Initial Setup

1) Using two culture tubes filled with 5 mL of RPMI media (see Materials and methods section), add 0.5 mL of well-mixed whole blood. Rinse pipet 3–4 times to expel all of the whole blood into the culture tube. (Use 3 culture tubes for blood from newborns.)
2) With a 1-mL syringe, slowly add 0.1 mL phytohemagglutinin (PhA-C; see Materials and Methods). Gently vortex mix or invert to ensure complete mixing.
3) Place cultures in a tray with a slight slope upwards towards cap (about an 18° angle). Loosen cap to allow CO_2 penetration.
4) Incubate 66 h at 37°C with 4.5% CO_2 and 90% humidity (as in a water-jacketed incubator).

b) Arresting Cells at Mitotic Metaphase

1) After 66 h of incubation, remove from incubator and gently resuspend to ensure a homogenous mixture.
2) Add 0.1 mL of working Velban (see Material and Methods) to each tube.
3) Mix each culture by gently swirling. Return to CO_2 incubator for 45 min. At this same time, place hypotonic solution sufficient for all cultures in the 37°C incubator to prewarm.

c) Harvesting

1) After 45 min in Velban, centrifuge the culture tubes for 10 min at 800× *g*.
2) Aspirate and discard the supernatant leaving 0.25–0.50 mL of liquid on top of the packed cells.
3) Resuspend cells by mixing on a vortex mixer at the lowest setting.
4) Slowly add 5–10 mL of prewarmed hypotonic to each tube while vortex mixing.
5) Gently invert tubes and place in a 37°C water bath for 45–50 min.
6) Centrifuge for 15 min at 800× *g*.
7) Aspirate and save supernatant, leaving 0.25–0.50 mL of liquid on packed cells.
8) Resuspend cells in remaining supernatant by gently mixing on vortex mixer at the lowest setting.
9) Using a Pasteur pipet with a rubber bulb while continuing to vortex mix, add the fixative solution (see Materials and Methods) *very slowly* to bring the total volume to 10 mL in each tube.

10) Allow to stand at room temperature for a minimum of 25 min.
11) Repeat steps #6–9, for a total of three changes of fixative.
12) After the third fixation, spin at 800× *g*, aspirate supernatant leaving 0.10–0.20 mL on layer of fixed white blood cells.
13) Add 0.10–1.0 mL of fresh fixative to suspension; the amount depends on the density of cell button.

d) Slide Preparation

1) Resuspend button by bubbling with a fresh Pasteur pipet. Be careful not to draw any liquid into the wide portion of pipet—this would make recovery of the mitoses very difficult.
2) Prepare single-well silanized slides, as described earlier. Dip slides in DEPC-distilled water. Freeze for one-half hour before use.
3) Place wet, prechilled slide onto bench while dropping 3–5 drops of specimen 4–5 feet (this may take practice) directly onto the slide. Allow slides to air-dry.
4) Add 2–3 drops of supernatant saved in step #7 above, which contains the red-blood-cell ghost membranes. Allow to air-dry once again.
5) Cure slides on a hot plate at 59°–60°C for 30–48 h.
6) Heat fix at 105°C for 5–10 s, then fix in 2% PFA for 1–2 hours. Next, treat with proteinase K (6 µg/mL) for 3–5 min. Inactivate the proteinase K by placing slides on 95°C heat block for 2 min. Wash in DEPC H_2O_2. Air-dry.
7) Perform *in situ* amplification as described above with one modification—use only 15–18 cycles instead of 30 cycles.

HYBRIDIZATION

Prepare a solution containing: 20 pmol/mL of the appropriate probe, 50% deionized formamide, 2× standard saline citrate (SSC) buffer, 10× Denhardt's solution, 0.1% sonicated salmon sperm DNA and 0.1% sodium dodecyl sulfate (SDS). The following is a convenient recipe:

Probe (^{33}P-labeled, biotinylated, or digoxigenin)	2 µL
Deionized formamide	50 µL
20× SSC*	10 µL
50× Denhardt's solution	20 µL
10 mg/mL ssDNA*	10 µL
10% SDS	1 µL
H_2O	7 µL
Total Volume:	100 µL

* See Materials and Methods section for preparation of 20× SSC buffer; the salmon sperm should be denatured at 94°C for 10 min before it is added to the hybridization buffer.

Note: 2% BSA can be added if one is observing nonspecific binding. For this purpose, one can add 10 μL of 20% BSA solution and reduce the amount of water.

1) Add 10 μL of hybridization mixture to each well and add coverslips.
2) Heat slides on a block at 95°C for 5 min.
3) Incubate slides at 48°C for 2 to 4 h in a humidified atmosphere.

Note: The optimal hybridization temperature is a function of the T_m of the probe. This must be calculated for each probe, as described earlier. However, the hybridization temperatures used should not be too high. If that occurs, then the formula for the hybridization solution should be modified and instead of 50% formamide, 40% formamide should be substituted (described further in the *in situ* hybridization section of Current Protocols in Molecular Biology).

Post-Hybridization Procedure for ^{33}P Probes

1) Wash slides in 2× SSC for 5 min.
2) Dip slides in 3× nuclear tract emulsion (Kodak NBT-2 diluted 1:1 with water; Eastman Kodak, Rochester, NY, USA).

Note: Steps 3–5 should be carried out in the dark.

3) Slides are air-dried, then incubated for 3–10 days in light-proof box with a drying agent.
4) Slides are developed for 3 min in Kodak D-19 developer, then rinsed in H_2O.
5) Slides fixed for 3 min in Kodak Unifax.
6) Slides are counterstained with May Grunewald Giemsa.

Post-Hybridization for Peroxidase-Based Color Development

1) Wash slides in 1× PBS twice for 5 min each time.
2) Add 10 μL of streptavidin-peroxidase complex (100 μg/mL in PBS pH 7.2). Gently apply the coverslips.
3) Incubate slides at 37°C for 1 h.
4) Remove coverslip, wash slides with 1× PBS twice for 5 min each time.
5) Add to each well 100 μL of 3′-amino-9-ethylene carbazole in the presence of 0.03% hydrogen peroxide in 50 mM acetate buffer (pH 5.0).
6) Incubate slides at 37°C for 10 min to develop the color—this step should be carried out in the dark. After this period, observe slides under a microscope. If color is not strong, develop for another 10 min.
7) Rinse slides with tap water and allow to dry.
8) Add 1 drop of 50% glycerol in PBS and apply the coverslips.
9) Analyze with optical microscope—positive cells will be stained a brownish red.

Post-Hybridization for Alkaline-Phosphatase-Based Color Development

1) After hybridization, remove coverslip, wash the slides with two soakings in 2× SSC at room temperature for 15 min.
2) Cover each well with 100 μL of blocking solution (see below), place the slides flat in a humid chamber at room temperature for 15 min.
3) Prepare a working conjugate solution by mixing 10 μL of streptavidin-alkaline phosphatase conjugate (40 μg/mL stock) with 90 μL of conjugate dilution buffer (see below) for each well.
4) Remove the blocking solution from each slide by touching absorbent paper to the edge of the slide.
5) Cover each well with 100 μL of working conjugate solution and incubate in the humid chamber at room temperature for 15 min. Do not allow the tissue to dry after adding the conjugate.
6) Wash slides by soaking in buffer A (see below) for 15 min at room temperature two times.
7) Wash slides once in alkaline substrate buffer (see below) at room temperature for 5 min.
8) Pre-warm 50 mL of alkaline-substrate buffer (see below) to 37°C in a Coplin jar. Just before adding the slides, add 200 μL NBT and 166 μL of BCIP (see below). Mix well.
9) Incubate slides in the NBT/BCIP solution at 37°C until the desired level of signal is achieved (usually from 10 min to 2 h). Check the color development periodically by removing a slide from the NBT/BCIP solution. Be careful not to allow the tissue to dry.
10) Stop the color development by rinsing the slides in several changes of deionized water. The tissue may now be counterstained.

Blocking solution: 50 mg/mL BSA (protein) in 100 mM Tris-HCl (pH 7.8), 150 mM NaCl and 0.2 mg/mL sodium azide.

Conjugate dilution buffer: 100 mM Tris-HCl, 150 mM $MgCl_2$, 10 mg/mL BSA and 0.2 mg/mL sodium azide.

Buffer A: 100 mM Tris-HCl (pH 7.5), 150 mM NaCl.

Alkaline substrate buffer: 100 mM Tris HCl (pH 9.5), 150 mM NaCl and 50 mM $MgCl_2$.

NBT (Nitro-blue-tetrazolium): 75 mg/mL NBT in 70% (vol/vol) dimethlyformamide.

BCIP (4-bromo-5-chloro-3-indolylphosphate): 50 mg/mL in 100% dimethylformamide.

Counterstaining and Mounting

1) If you are using red indicator color (like AEC), then use Gill's hematoxalin (Sigma Chemical); if using alkaline phosphatase as indicator color, then use

Nuclear Fast Red stain as counterstain (stain for 5 min at room temperature).

2) Rinse in several changes of tap water.
3) Dehydrate the sections through graded ethanol series (50%, 70%, 90%, 100% vol/vol for 1 min each).
4) Air-dry at room temperature.
5) For permanent mounting, a water-based medium such as CrystalMount™ or GelMount™ or an organic solvent-based medium such as Permount® (Fisher Scientific, Pittsburgh, PA, USA) can be used.
6) Apply one drop of mounting medium per 22-mm coverslip.
7) The slides may be viewed immediately, if you are careful not to disrupt the coverslip. The mounting medium will dry after sitting overnight at room temperature.

VALIDATION AND CONTROLS

The validity of *in situ* amplification-hybridization should be examined in every experiment. Attention here is especially necessary in laboratories first using the technique, because occasional technical pitfalls lie on the path to mastery. In an experienced laboratory, it is still necessary to continuously validate the procedure and to confirm the efficiency of amplification. To do this, we routinely run two or three sets of experiments in multi-welled slides simultaneously, for we must not only validate amplification, but also confirm the subsequent hybridization/detection steps as well.

In our lab, we frequently work with HIV. A common validation procedure we will conduct is to mix HIV-1-infected cells plus HIV-1-uninfected cells in a known proportion (i.e., 1:10, 1:100, etc.); then we confirm that the results are appropriately proportionate. To examine the efficiency of amplification, we use a cell line that carries a single copy or two copies of cloned HIV-1 virus, then look to see that proper amplification and hybridization has occurred.

In all amplification procedures, we use one slide as a control for nonspecific binding of the probe. Here we hybridize the amplified cells with an unrelated probe. We also use HLA-DQa probes and primers with human PBMC as positive controls, to check various parameters of our system.

In case one is using tissue sections, a cell suspension lacking the gene of interest can be used as a control. These cells can be added on top of the tissue section and then retrieved after the amplification procedure. The cell suspension can then be analyzed with the specific probe to see if the signal from the tissue leaked out and entered the cells floating above.

We suggest that researchers carefully design and employ appropriate positive and negative controls for their specific experiments. In the case of RT *in situ* amplification, one can use beta actin, HLA-DQa and other endogenous-abundant RNAs as the positive markers. Of course, one should always have a RT-negative control for RT *in situ* amplification, as well as DNase and non-DNase controls. Controls without *Taq* DNA polymerase plus primers and without primers should always be included.

MATERIALS AND METHODS

Slides

Heavy, Teflon-coated glass slides with three 10-, 12- or 14-mm-diameter wells for cell suspensions, or single oval wells for tissue sections, are available from Cel-Line Associates (Newfield, NJ, USA) or Erie Scientific (Portsmouth, NH, USA). These specific slide designs are particularly useful, for the Teflon coating serves to form distinct wells, each of which serves as a small reaction "chamber" when the coverslip is attached. Furthermore, the Teflon coating helps to keep the nail polish from entering the reaction chamber, and multiple wells allow for both a positive and negative control on the same slide.

Coplin Jars and Glass Staining Dishes

Suitable vessels for washing, fixing and staining 4–20 glass slides simultaneously are available from several vendors, including Fisher Scientific, VWR Scientific (West Chester, PA, USA) and Sigma Chemical.

2% Paraformaldehyde

1. Take 12 g paraformaldehyde (ultra pure Art. No. 4005; Merck, Darmstadt, Germany) and add to 600 mL 1× PBS.
2. Heat at 65°C for 10 min.
3. When the solution starts to clear, add 4 drops 10 *N* NaOH and stir.
4. Adjust to neutral pH and cool to room temperature.
5. Filter on Whatman's No. 1.

10× PBS Stock Solution pH 7.2–7.4

Dissolve 20.5 g NaH_2PO_4 H_2O and 179.9 g Na_2HPO_4 $7H_2O$ (or 95.5 g Na_2 HPO_4) in about 4 liters of double-distilled water. Adjust to the required pH (7.2–7.4). Add 701.3 g NaCl and make up to a total volume of 8 liters.

1× PBS

Dilute the stock 10× PBS at 1:10 ratio (i.e., 100 mL 10× PBS and 900 mL of water for 1 liter). Final concentration of buffer should be 0.01 M phosphate and 0.15 M NaCl.

0.3% Hydrogen Peroxide (H_2O) in PBS

Dilute stock 30% hydrogen peroxide (H_2O_2) at a 1:100 ratio in 1× PBS for a final concentration of 0.3% H_2O_2.

Proteinase K

Dissolve powder from Sigma Chemical in water to obtain 1 mg/mL concentration. Aliquot and store at -20°C.

Working solution: Dilute 1 mL of stock (1 mg/mL) into 150 mL of 1× PBS.

20× SSC

Dissolve 175.3 g of NaCl and 88.2 g of sodium citrate in 800 mL of water. Adjust the pH to 7.0 with a few drops of 10 *N* solution of NaOH. Adjust the volume to 1 liter with water. Sterilize by autoclaving.

2× SSC

Dilute 20× SSC: 100 mL of 20× SSC and 900 mL of water.

Solutions for Amplifying Chromosomal Bands

1. *RPMI 1640 Medium*: Per 100 mL, supplement with 15 mL fetal bovine serum, 1.5 Hepes buffer (IM), 0.1 mL gentamycin (0.1 mL heparin is optional).
2. *Velban*: Reconstitute vial with 10 mL sterile H_2O. From this solution, dilute 0.1 mL into 50 mL distilled H_2O. Store in refrigerator.
3. *EGTA hypotonic solution*: Dissolve 0.2 g EGTA powder, 3.0 g KCl, 4.8 g Hepes buffer into 1000 mL of distilled H_2O. Adjust pH to 7.4. Store in refrigerator prior to use. Prewarm to 37°C.
4. *PHA-C (Phytohemagglutinin)*: Reconstitute with 5 mL sterile H_2O. Aliquot into five 1-mL insulin syringes. Freeze four for later use and leave one in refrigerator.

Streptavidin Peroxidase

Dissolve powder from Sigma Chemical in PBS to make a stock of 1 mg/mL. Just before use, dilute stock solution in sterile PBS at a 1:30 ratio.

Color Solution

Dissolve one AEC (3-amino-9-ethyl-carbazole from Sigma Chemical) tablet in 2.5 mL of *N,N*-dimethylformamide. Store at 4°C in the dark.

Working Solution (make fresh before each use, keeping solution in the dark):

50 mM Acetate buffer, pH 5.0	5 mL
AEC solution	250 μL
30% H_2O_2	25 μL

Preparation of 50 mM Acetate Buffer pH 5.0

Add 74 mL of 0.2 *N* acetic acid (11.55 mL glacial acid/liter) and 176 mL of 0.2 M sodium acetate (27.2 g sodium acetate trihydrate in 1 liter) to 1 liter of deionized water and mix.

***In Situ* Hybridization Buffer (for 5 mL)**

Formamide	2.5 mL
Salmon sperm DNA (ssDNA) (10 mg/mL)*	0.5 mL
20× SSC	0.5 mL
50× Denhardt's solution	1.0 mL
10% SDS	0.05 mL

Water 0.450 mL

Total Volume:5.000 mL

* **Note:** Heat denature ssDNA at 94°C for 10 min before adding to the solution.

REFERENCES

1. **Bagasra, O.** Polymerase chain reaction *in situ*, p. 20-21. Amplifications (March 1990), Editorial note.
2. **Bagasra, O., S.P. Hauptman, H.W. Lischner, M. Sachs and R.J. Pomerantz.** 1992. Detection of HIV-1 provirus in mononuclear cells by *in situ* PCR. N. Engl. J. Med. *326*:1385-1391.
3. **Bagasra, O., T. Seshamma and R.J. Pomerantz.** 1993. Polymerase chain reaction *in situ*: Intracellular amplification and detection of HIV-1 proviral DNA and other specific genes. J. Immunol. Methods *158*:131-145.
4. **Bagasra, O. and R.J. Pomerantz.** 1993. HIV-1 provirus is demonstrated in peripheral blood monocytes in vivo: A study utilizing an *in situ* PCR. AIDS Res. Hum. Retroviruses *9*:69-76.
5. **Bagasra, O., T. Seshamma, J. Oakes and R.J. Pomerantz.** 1993. Frequency of cells positive for HIV-1 sequences assessed by *in situ* polymerase chain reaction. AIDS *7*:82-86.
6. **Bagasra, O., T. Seshamma, J. Oakes and R.J. Pomerantz.** 1993. High percentages of CD4-positive lymphocytes harbor the HIV-1 provirus in the blood of certain infected individuals. AIDS *7*:1419-1425.
7. **Bagasra, O., M.N. Qureshi, B. Joshi, I. Hewlett, C.E. Barr and D. Henrad.** 1994. High prevalence of HIV DNA and RNA and localization of HIV-proviral DNA in oral mucosal epithelial cells in saliva from HIV (+) subjects. Annual Meeting: United States and Canadian Academy of Pathology #742.
8. **Bagasra, O., T. Seshamma and R.J. Pomerantz.** 1993. *In situ* PCR: A powerful new methodology, p. 143-156. *In situ* Hybridization and Neurology. Oxford University Press, New York.
9. **Bagasra, O. and R.J. Pomerantz.** 1994. *In situ* PCR: Applications in the pathogenesis of diseases. Cell Vision *1*:13-16.
10. **Bagasra, O., H. Farzadegan, T. Seshamma, J. Oakes, A. Saah and R.J. Pomerantz.** Human immunodeficiency virus type 1 infection of sperm in vivo. AIDS *8*:1669-1674.
11. **Bagasra, O. and R.J. Pomerantz.** 1994. *In situ* polymerase chain reaction and HIV-1, p. 351-366. *In* R.J. Pomerantz (Ed.), Clinics of North America. W.B. Saunders Publishers, Philadelphia, PA.
12. **Bagasra, O. and R.J. Pomerantz.** Detection of HIV-1 in the brain tissue of individuals who died from AIDS, p. 339-358. *In* G. Sarkar (Ed.), PCR in Neuroscience. Academic Press, Orlando, FL.
13. **Bagasra, O., T. Seshamma, J.P. Pestaner and R.J. Pomerantz.** Detection of HIV-1 gene sequences in the brain tissues by *in situ* polymerase chain reaction, p. 251-266. *In* E. Majors and J.A. Levy (Eds.), Technical Advances in AIDS Research in the Nervous System. Plenum Press, New York (In press).
14. **Bagasra, O. and T. Seshamma.** Applications of *in situ* PCR methods in Molecular Biology. *In* J. Gu (Ed.), *In Situ* Polymerase Chain Reaction and Related Technology. Eaton Publishing, Natick, MA (In press).
15. **Bagasra, O., T. Seshamma, J. Hansen and R.J. Pomerantz.** 1995. *In Situ* PCR: A manual. *In* Current Protocols in Molecular Biology. John Wiley & Sons, New York.
16. **Bagasra, O., T. Seshamma, J. Hansen, L. Bobroski and R.J. Pomerantz.** 1994. Applications of *in situ* PCR methods in molecular biology: I. Details of methodology for general use. Cell Vision *1*:324-335.
17. **Cartun, R.W., J.F. Siles, L.M. Li, M.M. Berman and G.J. Nuovo.** 1994. Detection of hepatitis C virus infection in hepatectomy specimens using immunohistochemistry with reverse transcriptase (RT) *in situ* PCR confirmation. Cell Vision *1*:84.
18. **Cary, S.C., W. Waren, E. Anderson and S.J. Giovannoni.** 1993. Identification and localization of bacterial endosymbionts in hydrothermal Vent Taxa with symbiont-specific polymerase chain reaction amplification and *in situ* hybridization techniques. Mol. Marine Biol. Biotechnol. *2*:51-62.
19. **Chen, R.H. and S.V. Fuggle.** 1993. *In situ* cDNA polymerase chain reaction: A novel technique for detecting mRNA expression Am. J. Pathol. *143*:1527-1533.
20. **Chieu, K.-P., S.H. Cohen, D.W. Morris and G.W. Jordan.** 1992. Intracellular amplification of proviral DNA in tissue sections using the polymerase chain reaction. J. Histochem. Cytochem. *40*:333-341.

21. **Cirocco, R., M. Careno, C. Gomez, K. Zucker, V. Esquenazi and J. Miller.** 1994. Chimerism demonstrated on a cellular level by *in situ* PCR. Cell Vision *1*:84.
22. **Don, R.H., P.T. Cox, B.J. Wainwright, K. Baker and J.S. Matiick.** 1994. Touchdown PCR to circumvent spurious priming during gene amplification. Nucleic Acids Res. *19*:4008-4010.
23. **Embleton, M.J., G. Gorochov, P.T. Jones and G. Winter.** 1992. In-cell PCR from mRNA amplifying and linking the rearranged immunoglobulin heavy and light chain V-genes within single cells. Nucleic Acids Res. *20*:3831-3837.
24. **Embretson, J., M. Zupanic, T. Beneke, M. Till, S. Wolinsky, J.L. Ribas, A. Burke and A.T. Haase.** 1993. Analysis of human immunodeficiency virus-infected tissues by amplification and *in situ* hybridization reveals latent and permissive infections at single-cell resolution. Proc. Natl. Acad. Sci. USA *90*:357-361.
25. **Embretson, J., M. Zupancic, J.L. Ribas, A. Burke, P. Racz, K. Tenner-Racz and A.T. Haase.** 1993. Massive covert infection of helper T lymphocytes and macrophages by HIV during the incubation period of AIDS. Nature *62*:359-362.
26. **Gingeras, T.R., P. Prodanovich, T. Latimer, J.C. Guatelli, D.D. Richman and KJ. Baninger.** 1991. Use of self-sustained sequence replication amplification reaction to analyze and detect mutations in zidovudine-resistant human immunodeficiency virus. J. Infect. Dis. *164*:1066-1074.
27. **Gingeras, T.R., D.D. Richman, D.Y. Kwoh and J.C. Guatelli.** 1990. Methodologies for in vitro nucleic acid amplification and their applications. Vet. Microbiol. *24*:235-251.
28. **Gingeras, T.R., K.M. Whitfield and D.Y. Kwoh.** 1990. Unique features of the self-sustained sequence replication (3SR) reaction in the in vitro amplification of nucleic acids. Ann. Biol. Clin. (Paris) *4*:498-501.
29. **Gosden, J. and D. Hanratty.** 1993. PCR *in situ*: A rapid alternative to *in situ* hybridization for mapping short, low copy number sequences without isotopes. BioTechniques *5*:78-80.
30. **Greer, C.E., J.K. Lund and M.M. Manos.** 1991. PCR amplification from paraffin-embedded tissues: Recommendations on fixatives for long-term storage and prospective studies. PCR Method Appl. *95*:117-124.
31. **Greer, C.E., S.L. Peterson, N.B. Kiviat and M.M. Manos.** 1991. PCR amplification from paraffin-embedded tissues: Effects of fixative and fixative times. Am. J. Clin. Pathol. 95:117-124.
32. **Gressens, P., C. Langston and J.R. Martin.** 1994. *In situ* PCR localization of herpes simplex virus DNA sequences in disseminated neonatal herpes encephalitis. J. Neuropathol. Exp. Neurol. *53*:469-482.
33. **Gressens, P. and J.R. Martin.** 1994. HSV-2 DNA persistence in astrocytes of the trigeminal root entry zone: Double labeling by *in situ* PCR and immunohistochemistry. J. Neuropathol. Exp. Neurol. *53*:127-135.
34. **Guatelli, J.C., K.M. Whitfield, D.Y. Kwoh, K.J. Barringer, D.D. Richman and T.R. Gingeras.** 1994. Isothermal, in vitro amplification of nucleic acids by a multienzyme reaction modeled after retroviral replication. Proc. Natl. Acad. Sci. USA *7*:1874-1878.
35. **Haase, A.T., E.F. Retzel and K.A. Staskus.** 1990. Amplification and detection of lentiviral DNA inside cells. Proc. Natl. Acad. Sci. USA *37*:4971-4975.
36. **Heniford, B.W., A. Shum-Siu, M. Leonberger and F.J. Hendler.** 1993. Variation in cellular EGF receptor mRNA expression demonstrated by *in situ* reverse transcription polymerase chain reaction. Nucleic Acids Res. *21*:3159-3166.
37. **Hiort, O., G. Klauber, M. Cendron, G.H. Sinnecker, L. Keim, E. Schwinger, H.J. Wolfe and D.W. Yandell.** 1994. Molecular characterization of the androgen receptor gene in boys with hypospadias. Eur. J. Pediatr. *153*:317-321.
38. **Hacker, G.W., I. Zehbe, C. Hauser-Kronberger, J. Gu, A.-H. Graf and O. Dietze.** 1994. *In situ* detection of DNA and mRNA sequences by immunogold-silver staining (IGSS). Cell Vision *1*:30-37.
39. **Heniford, B.W., A. Shum-Siu, M. Leonberger and F.J. Hendler.** 1993. Variation in cellular EGF receptor mRNA expression demonstrated by *in situ* reverse transcriptase polymerase chain reaction. Nucleic Acids Res. *21*:3159-3166.
40. **Hsu, T.C., O. Bagasra, T. Seshamma and P.N. Walsh.** 1994. Platelet factor XI mRNA amplified from human platelets by reverse transcriptase polymerase chain reaction and detected by *in situ* amplification and hybridization. FASEB J. *8*:1375.
41. **Li, H.H., U.B. Gyllensten, X.F. Cui, R.K. Saiki, H.A. Erlich and N. Arnheim.** 1988. Amplification and analysis of DNA sequences in single human sperm and diploid cells. Nature *335*:414-417.
42. **Isaacson, S.H., D.M. Asher, C.J. Gibbs and D.C. Gajdusek.** 1994. *In situ* RT-PCR amplification in archival brain tissue. Cell Vision *1*:84.

43. **Kawasaki, E., R. Saiki and H. Erlich.** 1989. PCR Technology. Principles and Applications for DNA Amplification. H.A. Ehrlich (Ed.). Stockton Press, New York.
44. **Komminoth, P., A.A. Long, R. Ray and H.J. Wolfe.** 1992. *In situ* polymerase chain reaction detection of viral DNA. Single copy genes and gene rearrangements in cell suspensions and cytospins. Diagn. Mol. Pathol. 1:85-97.
45. **Komminoth, P., F.B. Merk, I. Leav, H.J. Wolfe and J. Roth.** 1992. Comparison of 32S and digoxigenin-labeled RNA and oligonucleotide probes for *in situ* hybridization expression of mRNA of the seminal vesicle secretion protein 11 and androgen receptor genes in the rat prostate. Histochemistry *93*:217-228.
46. **Kuwata, S.** 1992. Application of PCR and RT-PCR method to molecular biology study in nephrology. Nippon Rinsho *50*:2868-2873.
47. **Larzul, D., F. Guigue, J.J. Sninsky, D.H. Mach, C. Brechot and J.L. Guesdaon.** 1988. Detection of hepatitis B virus sequences in serum by using in vitro enzymatic amplification J. Virol. Methods *20*:227-237.
48. **Long, A.A., P. Komminoth, E. Lee and H.J. Wolfe.** 1993. Comparison of indirect and direct *in situ* polymerase chain reaction in cell preparations and tissue sections. Detection of viral DNA gene rearrangements and chromosomal translocations. Histochemistry *99*:151-162.
49. **Lundberg, K.S., D.D. Shoemaker, M.W. Adams, J.M. Short, J.A. Sorge and E.J. Mathur.** 1991. High fidelity amplification using a thermostable DNA polymerase isolated from Pyrococcus furiosus. Gene *1*:81-86.
50. **Mahalingham, R., S. Kido, M. Wellish, R. Cohrs and D.H. Gilden.** *In situ* polymerase chain reaction detection of varicella zoster virus in infected cells in culture (In press).
51. **Mankowski, J.L., J.P. Spelman, H.G. Reesetar, J.D. Strandberg, J. Laterra, D.L. Carter, J.E. Clements and M.C. Zink.** 1994. Neurovirulent Simian immunodeficiency virus replicates productively in endothelial cells of the central nervous system in vivo and in vitro. J. Virol. *68*:8202-8208.
52. **Mehta, A., J. Maggioncalda, O. Bagasra, S. Thikkavarapu, P. Saikumari and T. Block.** 1994. Detection of herpes simplex sequences in the trigeminal ganglia of latently infected mice by *in situ* PCR method. Cell Vision *1*:110-115.
53. **Mehta, A., J. Maggioncalda, O. Bagasra, S. Thikkavarapu, P. Saikumari, F.W. Nigel and T. Block.** 1994. *In situ* PCR and RNA hybridization detection of Herpes Simplex virus sequences in trigeminal ganlia of latently infected mice. Virology *206*:633-640.
54. **Mitchell, W.J., P. Gressens, J.R. Martin and R. DeSanto.** 1994. Herpes simplex virus type 1 DNA persistence, progressive disease and transgenic immediate early gene promoter activity in chronic corneal infections in mice. J. Gen. Virol. *75*:1201-1210.
55. **Murray, G.I.** 1993. *In situ* PCR. J. Pathol. *169*:187-188.
56. **Nuovo, G.J.** 1990. Human papillomavirus DNA in genital tract lesions histologically negative for condylomata. Analysis by *in situ*, Southern blot hybridization and the polymerase chain reaction. Am. J. Surg. Pathol. *14*:643-651.
57. **Nuovo, G.J.** 1994. Questioning *in situ* PCR. *In situ* cDNA polymerase chain reaction: A novel technique for detecting mRNA expression. Am. J. Pathol. 145:741.
58. **Nuovo, G.J., J. Becker, M. Margiotta, P. MacConnell, S. Comite and H. Hochman.** 1992. Histological distribution of polymerase chain reaction-amplified human papillomavirus 6 and 11 DNA in penile lesions. Am. J. Surg. Pathol. *16*:269-275.
59. **Nuovo, G.J., J. Becker, A. Simsir, M. Margiotta, G. Khalife and M. Shevchuk.** 1994. HIV-1 nucleic acids localize to the spermatogonia and their progeny. A study by polymerase chain reaction *in situ* hybridization. Am. J. Pathol. *144*:1142-1148.
60. **Nuovo, G.J., M.M. Darfler, C.C. Impraim and S.E. Bromley.** 1991. Occurrence of multiple types of human papillomavirus in genital tract lesions. Analysis by *in situ* hybridization and the polymerase chain reaction. Am. J. Pathol. *138*:53-58.
61. **Nuovo, G.J., P. Delvenne, P. MacConnell, E. Chalas, C. Neto and M.J. Mann.** 1991. Correlation of histology and detection of human papillomavirus DNA in vulvar cancers. Gynecol. Oncol. *43*:275-280.
62. **Nuovo, G.J., A. Forde, P. MacConnell and R. Fahrenwald.** 1993. *In situ* detection of PCR-amplified HIV-1 nucleic acids and tumor necrosis factor cDNA in cervical tissues. Am. J. Pathol. *143*:40-48.
63. **Nuovo, G.J., F. Gallery, R. Hom, P. MacConnell and W. Bloch.** *1993*. Importance of different variables for enhancing in situ detection of PCR-amplified DNA. PCR Meth. Appl. *2*:305-312.

64.**Nuovo, G.J., F. Gallery and P. MacConnell.** 1992. Detection of amplified HPV 6 and 11 DNA in vulvar lesions by hot start PCR *in situ* hybridization. Mod. Pathol. *5*:444-448.
65.**Nuovo, G.J., F. Gallery, P. MacConnell and A. Braun.** 1994. *In situ* detection of polymerase chain reaction-amplified HIV-1 nucleic acids and tumor necrosis factor-alpha RNA in the central nervous system. Am. J. Pathol. *144*:659-666.
66.**Nuovo, G.J., F. Gallery, P. MacConnell, J. Becker and W. Bloch.** 1991. An improved technique for the *in situ* detection of DNA after polymerase chain reaction amplification. Am. J. Pathol. *139*:1239-1244.
67.**Nuovo, G.J., G.A. Gorgone, P. MacConnell, M. Margiotta and P.D. Gorevic.** 1992. *In situ* localization of PCR-amplified human and viral cDNA. PCR Meth. Appl. *2*:117-123.
68.**Nuovo, G.J., H.A. Hocman, Y.D. Eliezri, D. Lastarria, S.L. Comite and D.N. Silvers.** 1990. Detection of human papillomavirus DNA in penile lesions histologically negative for condylomata. Analysis by *in situ* hybridization and the polymerase chain reaction. Am. J. Surg. Pathol. *14*:829-836.
69.**Nuovo, G.J., K. Lidonnici, P. MacConnell and B. Lane.** 1993. Intracellular localization of polymerase chain reaction (PCR)-amplified hepatitis C cDNA. Am. J. Surg. Pathol. *17*:683-690.
70.**Nuovo, G.J., P. MacConnell, A. Forde and P. Delvenne.** 1991. Detection of human papillomavirus DNA in formalin-fixed tissues by *in situ* hybridization after amplification by polymerase chain reaction. J. Pathol. *139*:847-854.
71.**Nuovo, G.J., M. Margiotta, P. MacConnell and J. Becker.** 1992. Rapid *in situ* detection RT-PCR-amplified HIV-1 DNA. Diagn. Mol. Pathol. *1*:98-102.
72.**Nuovo, G.J., J. Moritz, L.L. Walsh, P. MacConnell and J. Koulos.** 1992. Predictive value of human papillomavirus DNA detection by filter hybridization and polymerase chain reaction in women with negative results of colposcopic examination. Am. J. Clin. Pathol. *98*:489-492.
73.**Nuovo, M., G.J. Nuovo, J. Becker, F. Gallery, P. Delvenne and P.B. Kane.** 1993. Correlation of viral infection, histology, and mortality in immuno-compromised patients with pneumonia. Analysis by *in situ* hybridization and the polymerase chain reaction. Diagn. Mol. Pathol. *2*:0-9.
74.**Nuovo, M., G.J. Nuovo, P. MacConnell, A. Forde and G.C. Steiner.** 1992. *In situ* analysis of Paget's disease of bone for measles-specific PCR-amplified cDNA. Diagn. Mol. Pathol. *1*:256-265.
75.**Nuovo, G.J.** 1994. PCR *In situ* Hybridization. Protocols and Applications, 2nd Edition. Raven Press, NY.
76.**O'Leary, J.J., G. Browne, R.J. Landers, M. Crowley, I.B. Healy, J.T. Street, A.M. Pollock, J. Murphy, M.I. Johnson and F.A. Lewis.** 1994. The importance of fixation procedures on DNA template and its suitability for solution-phase polymerase chain reaction and PCR and *in situ* hybridization. Histochem. J. *26*:337-346.
77.**Patel, V.G., A. Shum-Siu, B.W. Heniford, T.J. Wieman and F.J. Hendler.** Detection of epidermal growth factor receptor mRNA in tissue sections from biopsy specimens using *in situ* polymerase chain reaction Am. J. Pathol. *144*:7-14.
78.**Patterson, B.K., M. Till, P. Otto, C. Goolsby, M.R. Furtado, L.J. McBride and S.M. Wolinsky.** 1993. Detection of HIV-1 DNA and messenger RNA in individual cells by PCR-driven *in situ* hybridization and flow cytometry. Science *260*:976-979.
79.**Pestaner, J.P., M. Bibbo, L. Bobroski, T. Seshamma and O. Bagasra.** 1994. Potential of *in situ* polymerase chain reaction in diagnostic cytology. Acta Cytologia *38*:676-680.
80.**Pestaner, J.P., M. Bibbo, L. Bobroski, T. Seshamma and O. Bagasra.** 1994. Surfactant protein A mRNA expression utilizing the reverse transcription *in situ* PCR for metastatic adenocarcinoma. Cell Vision *1*:290-293.
81.**Qureshi, M.N., C.E. Barr, T. Seshamma, R.J. Pomerantz and O. Bagasra.** Localization of HIV-1 proviral DNA in oral mucosal epithelial cells. J. Infect. Dis. *171*:190-193.
82.**Qureshi, M.N., O. Bagasra, B. Joshi, I. Howlett, C.E. Barr and D. Henrad.** 1994. High prevalence of HIV DNA and RNA and localization of HIV-provirus DNA in oral mucosal epithelial cells in saliva from HIV+ subjects. Lab. Invest. *70*:127A (#742).
83.**Ray, R., P. Komminoth, M. Machado and H.J. Wolfe.** 1991. Combined polymerase chain reaction and *in situ* hybridization for the detection of single copy genes and viral genomic sequences in intact cells. Mod. Pathol. *4*:124A.
84.**Saito, H., A. Nishikawa, J. Gu, Y. Ihara, H. Soejima, Y. Wada, C. Sekiya, N. Niikawa and N. Taniguchi.** 1994. cDNA cloning and chromosomal mapping of human N-acetylglucosaminyl transferase V+. Biochem. Biophys. Res. Comm. *198*:318-327.
85.**Sällström, J.F., I. Zehbe, M. Alemi and E. Wilander.** 1993. Pitfalls of *in situ* polymerase chain reaction (PCR) using direct incorporation of labelled nucleotides. Anticancer Res. *13*:1153.
86.**Schwartz, D., U. Sharma, M. Busch, K. Weinhold, J. Lieberman, D. Birx, H. Farzedagen,**

J. Margolick, T. Quinn, B. Davis, S. Leitman, O. Bagasra, R.J. Pomerantz and R. Viscidi. 1994. Absence of recoverable infectious virus and unique immune responses in an asymptomatic HIV+ long term survivor. AIDS Res. Hum. Retroviruses *10*:1703-1711.

87. **Spann, W., K. Pachmann, H. Zabnienska, A. Pielmeier and B. Emmerich.** 1991. *In situ* amplification of single copy gene segments in individual cells by the polymerase chain reaction. Infection *19*:242-244.

88. **Staskus, K.A., L. Couch, P. Bitterman, E.F. Retzel, M. Zupancic, J. List and A.T. Haase.** 1991. *in situ* amplification of visna virus DNA in tissue sections reveals a reservoir of latently infected cells. Microb. Pathog. *11*:67-76.

89. **Stork, P., M. Loda, S. Bosari, B. Wiley, K. Poppenhusen and H.J. Wolfe.** 1992. Detection of K-ras mutations in pancreatic and hepatic neoplasms by non-isotopic mismatched polymerase chain reaction. Oncogene *6*:857-862.

90. **Staecker, H., M. Cammer, R. Rubenstein and T. Van De Water.** 1994. A procedure for RT-PCR amplification of mRNAs on histological specimens. BioTechniques *16*:76-80.

91. **Staskus, K.A., J. Embretson, E.F. Retzel, A.T. Haase and P. Bitterman.** 1994. Wild DNA and wild host cell mRNA *in situ*, p. 55-66. *In* K. Mullis, F. Ferre and R. Gibbs (Eds.), PCR: The Polymerase Chain Reaction. Birkhauser, Boston.

92. **Sukpanichnant, S., C.L. Vnencak-Jones and T.L. McCurley.** 1993. Detection of clonal immunoglobulin in heavy chain gene rearrangements by polymerase chain reaction in scrapings from archival hematoxylin and eosin-stained histologic sections: Implications for molecular genetic studies of focal pathologic lesions. Diagn. Mol. Pathol. *2*:168-176.

93. **Tsongalis, G.J., A.H. McPhail, R.D. Lodge-Rigal, J.F. Chapman and L.M. Silverman.** 1994. Localized *in situ* amplification (LISA): A novel approach to *in situ* PCR. Clin. Chem. *40*:381-384.

94. **Walter, M.J., T.J. Lehky, C.H. Fox and S. Jacobson.** 1994. *In situ* PCR for the detection of HTLV-1 in HAM/TSP patients. Ann. NY Acad. Sci. *724*:404-413.

95. **Walboomers, J.M.M., W.J.G. Melchers, H. Mullink, C.L.M. Meijer, A. Struyk, W.G.J. Quint, J. van der Noorda and J. ter Schegget.** 1988. Sensitivity of *in situ* detection with biotinylated probes of human papillomavirus type 16 DNA in frozen tissue sections of squamous cell carcinoma of the cervix. Am. J. Pathol. *139*:587-594

96. **Winslow, B.J., R.J. Pomerantz, O. Bagasra and D. Trono.** 1993. HIV-1 latency due to the site of proviral integration. Virology *196*:849-854.

97. **Wolfe, H.J., D. Ross and B. Wolfe.** 1990. Detection of infectious agents by molecular methods at the cellular level. Verhandlungen der Deutschen Gesellschaft fur Pathologie *74*:295-300.

98. **Yap, E.P.H. and J.O'D. McGee.** 1991. Slide PCR: DNA amplification from cell samples on microscopic glass slides. Nucleic Acids Res. *19*:1-5.

99. **Yin, J., M.G. Kaplitt and D.W. Pfaff.** 1994. *In situ* PCR and in vivo detection of foreign gene expression in rat brain. Cell Vision *1*:58-59.

100. **Zehbe, I., G.W. Hacker, E. Rylander, J. Sällström and E. Wilander.** 1992. Detection of single HPV copies in SiHa cells by *in situ* polymerase chain reaction (*in situ* PCR) combined with immunoperoxidase and immunogold-silver staining (IGSS) techniques. Anticancer Res. *12*:2165-2168.

101. **Zehbe, I., G.W. Hacker, J.F. Sällström, W.H. Muss, C. Hauser-Kronberger, E. Rylander and E. Wilander.** 1994. PCR *in situ* hybridization (PISH) and *in situ* self-sustained sequence replication-based amplification (*in situ* 3SR). Cell Vision *1*:46-47.

102. **Zehbe, I., G.W. Hacker, J.F. Sällström, E. Rylander and E. Wilander.** 1994. Self-sustained sequence replication-based amplification (3SR) for the *in situ* detection of mRNA in cultured cells. Cell Vision *1*:20-24.

103. **Zehbe, I., J.F. Sällström, G.W. Hacker, C. Hauser-Kronberger, E. Rylander and E. Wilander.** 1994. Indirect and direct *in situ* PCR for the detection of human papillomavirus. An evaluation of two methods and a double staining technique. Cell Vision *1*:163-168.

104. **Zevallos, E., E. Bard, V. Anderson and J. Gu.** 1994. An *in situ* (ISPCR) study of HIV- 1 infection of lymphoid tissues and peripheral lymphocytes. Cell Vision *1*:87.

105. **Zevallos, E., E. Bard, V. Anderson and J. Gu.** 1994. Detection of HIV-1 sequences in placentas of HIV-infected mothers by *in situ* PCR. Cell Vision *1*:116-121.

Address correspondence to Omar Bagasra, The Dorrance H. Hamilton Laboratories, Section of Molecular Retrovirology, Division of Infectious Diseases, Department of Medicine, Thomas Jefferson University, Philadelphia, PA 19102, USA.

In Situ PCR for the Detection of Human Papillomavirus in Cells and Tissue Sections

Ingeborg Zehbe[1], Jan F. Sällström[1], Gerhard W. Hacker[2], Eva Rylander[3] and Erik Wilander[1]

Departments of [1]Pathology and [3]Gynecology/Obstetrics, University Hospital, Uppsala, Sweden; [2]Institute of Pathological Anatomy, Immunohistochemistry and Biochemistry Unit, Landeskrankenanstalten Salzburg, Salzburg, Austria

SUMMARY

In situ *hybridization (ISH) with labeled nucleic acid probes has become a valuable tool for the detection of human papillomavirus (HPV), as it allows direct correlation of viral infection and morphological diagnosis. This method, however, is limited to the detection of 10–50 HPV DNA copies per cell, which was demonstrated on HeLa cells. We present here a more sensitive* in situ *method, PCR* in situ *hybridization (PISH), so-called indirect* in situ *PCR, which is able to detect 1–2 HPV DNA copies in SiHa cells. These stain negative with conventional ISH and positive with PISH. We have also evaluated a rapid* in situ *PCR method, direct* in situ *PCR, which is based on direct incorporation of labeled nucleotides during the polymerization process.*

INTRODUCTION

In situ hybridization (ISH) with labeled nucleic acid probes for the detection of human papillomavirus (HPV) has become an adjunct to histopathological diagnosis. This method, however, is limited to the detection of 10–50 HPV copies per cell, since HeLa cells (10–50 HPV 18 copies per cell) stain positive and SiHa cells (1–2 HPV 16 copies per cell) do not (23–24). In high grade squamous intraepithelial lesions (SILs) and invasive cancers, the HPV-copy number is often below this number (19). The polymerase chain reaction (PCR) *in vitro* is probably the most sensitive method for viral detection. For this reason, a method that combines the sensitivity of PCR and the possibility to relate the molecular biological diagnosis to morphology has been developed.

Nuovo et al. (13) were the first to report on PCR *in situ* hybridization (indirect *in situ* PCR) for the detection of HPV in tissue sections. At first they applied a modified version of Haase et al. (7,21) using multiple, overlapping primers. This, so Haase argued, would result in long amplicons, 900–1200 bp, and thus prevent leakage of the amplicons from the nuclei. Reasoning in the same way, Chiu et al. (4) developed a system employing complementary primer tails. Later, Nuovo et al.

(14,15) exchanged his assay for just one primer pair and "hot start" PCR and reported successful experiments. To make it still easier, Nuovo et al. (16) and Spann et al. (20) developed a rapid *in situ* PCR method (direct *in situ* PCR), which is based on direct incorporation of labeled nucleotides during the polymerization process. The difference compared to PCR *in situ* hybridization is that no nucleic acid probe is required. The only step needed is a conjugated antiserum to the label used for the incorporated nucleotides and an appropriate detection step for the visualization of positive results in the microscope. Meanwhile, various protocols have been published using either indirect or direct *in situ* PCR with modifications for different applications (1–3,6,10–12,25,26).

We have also reported that direct *in situ* PCR, according to Nuovo, may be applied with success on SiHa cells (25). However, when we performed a comparable assay on clinical material, e.g., anogenital tissue sections, we often noticed unspecific staining when omitting the primers in the negative controls (17,18). Similar results have also been reported by Long et al. (12).

We present here an improved indirect *in situ* PCR method, PCR *in situ* hybridization, designated PISH, using one primer pair and the "hot start" technique with subsequent hybridization to a biotin-labeled *pan*-probe for HPV. Our improvements included the use of a thermal cycler specifically equipped with a solid heating block for glass slides and a special setup using a PAP pen and mineral oil coating with two coverslips applied consecutively directly after the "hot start" for better reproducibility of the PCR. This assay works well on whole cell preparations, e.g., cytospins and imprints, but may need further optimizing for tissue sections. We also demonstrate false positivity on clinical biopsies, the outcome of direct *in situ* PCR. (As mentioned above, a probe is not required for this method as labeled nucleotides are built in directly during PCR.)

MATERIALS AND METHODS

Indirect and Direct *In Situ* PCR in Cells and Tissue Sections

Cultured SiHa cells as well as imprints and biopsies from anogenital lesions were fixed in 4% neutral phosphate-buffered formaldehyde overnight at room temperature. The SiHa cells were processed as cytospins on aminopropyl-triethoxysilane (APES)-coated slides and rinsed in 70%, 95% and 99% ethanol and air-dried. Sections were deparaffinized using xylene, transferred to 99% ethanol and air-dried. A predigestion of 0.1 mg proteinase K (Code No. 1413 783; Boehringer Mannheim, Mannheim, Germany)/mL sodium chloride sodium phosphate (SSPE), pH 7.4 for 5–30 min at 37°C depending on fixation time has been found to give reproducible results. If detection with peroxidase is performed, the endogenous peroxidase must be quenched with aqueous 6% H_2O_2 for 10 min at room temperature.

For the PCR process, a thermal cycler, PHC-3™ (Techne, Cambridge, UK) equipped with a flat, smooth thermo-block holding up to 5 glass slides, was utilized. For pipetting, aerosol-resistant and DNase-free pipet tips were applied. To prevent evaporation during the polymerization process, the dried samples were encircled with a PAP pen (SCI Science Services, Munich, FRG).

The amplification mixture consisted of 1.0 μM of each HPV consensus primer (20), 200 μM of each dNTP, 4.5 mM $MgCl_2$, 50 mM KCl, 10 mM Tris-HCl pH 8.3, 7.5 U AmpliTaq DNA Polymerase (Perkin-Elmer, Norwalk, CT, USA) per 50 μL amplification mixture and glass-distilled water. For direct *in situ* PCR, 10–20 μM digoxigenin-labeled dUTP (Boehringer Mannheim) was added. All reagents should be aliquoted and stored at -20°C.

"Hot start" PCR. The slides were placed on the thermo-block of the thermocycler and preheated to 82°C. Meanwhile, the amplification mixture was equally heated on an Eppendorf thermomixer (Madison, WI, USA), the AmpliTaq added and 10–25 mL of mixture were rapidly pipetted onto the samples, coverslipped within the wax circle and overlaid with 100–200 mL of mineral oil. An additional, larger coverslip spread the oil evenly and prevented evaporation. Thermocycling was performed according to the following profile: Initial denaturation for 3 min at 95°C and 25 cycles of annealing for 2 min at 55°C, extension for 1 min at 72°C and denaturation for 1 min at 95°C. Subsequently, the mineral oil was removed by placing the slides in xylene. This was followed by immersion in 99% ethanol and air-drying.

Detection method for indirect *in situ* PCR, PCR *in situ* hybridization (PISH). The amplified viral DNA was hybridized to a biotinylated HPV DNA *pan*-probe (4) essentially as described earlier and detected by enzymatic methods such as direct immunoperoxidase (IMP) or alkaline phosphatase (AP) (23,24) as well as immunogold silver-staining (IGSS) (8,9). Briefly, hybridization was carried out in a humid chamber at 37°C overnight. Post-hybridization was performed in 30% formamide/2× SSPE (final concentration) at 37°C for 10 min, followed by enzymatic detection with biotinylated horseradish peroxidase or AP bound to streptavidin (DAKO, Copenhagen, Denmark) 1:100 at 37°C for 30 min. Visualization of the enzyme label was achieved with diaminobenzidine-tetrahydrochloride (DAB)-H_2O_2 (peroxidase) at room temperature for 5–15 min and slight counterstaining with hematoxylin for 30 s or NBT/BCIP (alkaline phosphatase) at 37°C for 60 min and counterstaining with eosin for 1–5 min. For the development with IGSS, we used 1 nm and 5 nm gold-labeled anti-biotin antibodies (Amersham, London, UK) at a dilution of 1:50 each in Tris-buffered saline (TBS), pH 7.6, and 0.4% cold water fish gelatin at 4°C overnight. These were subsequently amplified by silver acetate autometallography to easily yield visible signals in the light microscope (8).

Detection method for direct *in situ* PCR. The digoxigenin-labeled nucleotides that had been built into the amplicons during the PCR process were detected as previously described (25). Briefly, peroxidase-labeled sheep anti-digoxigenin antibodies were applied, detected with DAB-H_2O_2 for 5–15 min at room temperature and counterstained with hematoxylin for 30 s.

Controls. Negative controls included the omission of the *Taq* DNA Polymerase and the primers. Known HPV-negative cell lines, such as human foreskin fibroblasts, served also as negative controls. The positive controls were SiHa cells, assuring that PCR worked satisfactorily. It is advisable to test all specimens with *in vitro* PCR as a further control of staining and reaction specificity.

RESULTS

Indirect *In Situ* PCR, PCR *In Situ* Hybridization (PISH)

All controls used were satisfactory. The PISH protocol with IMP resulted in light to dark brown nucleic staining in SiHa cells. Light to dark blue signals were visible using AP. The IGSS detection led to black staining of the HPV-infected nuclei (Figures 1 and 2). In cases where the copy number of viral genomes was above the detection limit of ISH, the ISH detection itself gave positive reactions too; however, the number of positive cells and the staining intensity were often greater in the PISH samples than in those treated with ISH alone. We sometimes noticed considerable background in tissue sections manifesting itself in unspecific cytoplasmic staining. We suppose this is due to leaking amplicons from the nuclei during the PCR process.

Direct *In Situ* PCR

This method did not meet all our requirements regarding controls. The SiHa cells stained positive after direct *in situ* PCR, including all reagents, but negative when one of the following was omitted: the *Taq* DNA Polymerase, the $MgCl_2$ or the primers. However, many of our anogenital biopsies showed unspecific, positive results specifically localized to the nuclei. These appeared in the upper layers of the epithelium, often in cells with pycnotic nuclei, and could easily be mistaken for a specific signal. This phenomenon was noticed in the negative controls when primers were left out, which proves the false positivity of these samples.

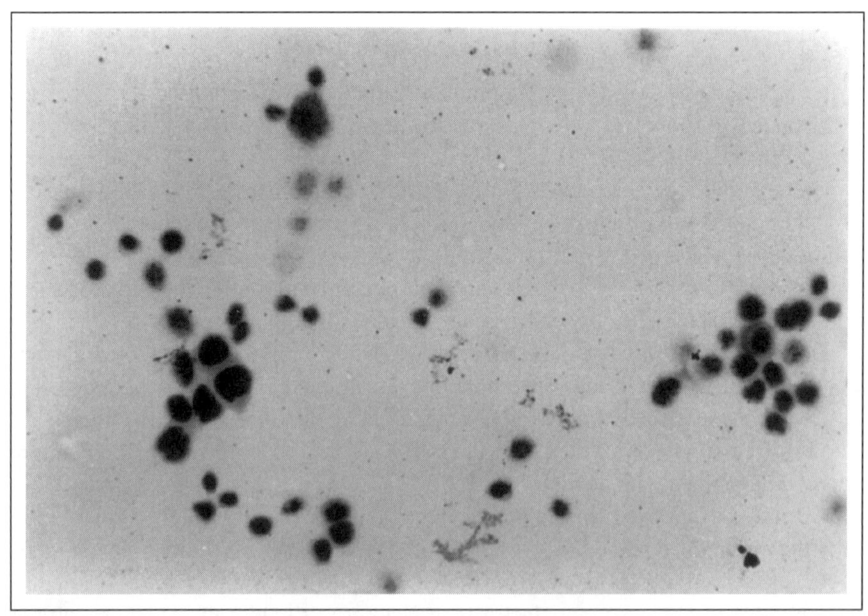

Figure 1. SiHa cells treated with PISH-IGSS. Positive results manifested themselves as black staining in the nuclei.

DISCUSSION

The PISH technique described, although still under revision, is more sensitive than ISH alone. ISH has a detection limit of 10–50 viral copies per cell, which was demonstrated on HeLa cells (23,24). PISH can detect 1–2 viral copies per cell, since SiHa cells stained positive with this technique but negative with ISH. Thus, PISH works well on whole cell preparations.

Regarding tissue sections, protocols are still under improvement; its biggest problem is leaking amplicons from the nuclei in some cases. This was less evident in cell preparations because their nuclei are not opened or "sliced" as in biopsy material. The main reason for losing amplicons may be the reiterated heating, due to DNA denaturating steps during PCR, which subjects the newly synthesized fragments to increased diffusion pressure. In ultrastructural micrographs of SiHa cell monolayers, treated with PISH-IGSS as a pre-embedding procedure, the amplicons not only accumulated at the nucleus but also at the cytoplasmic border (unpublished observations).

In "thicker" sections, e.g., ≥10 μm, some complete nuclei will be preserved. This is probably the closest one can get to whole cell preparations and may help to better retain amplified DNA to the nucleus. Application of primers with complementary tails as described by Chiu et al. (5) might also improve conditions for PISH on sections. Haase et al.'s approach to overcome this problem with multiple primers requires rather long preserved DNA regions. This may work well on cell preparations and prospective material with appropriate fixative but could present problems for archival, formalin-fixed biopsies—the major source of samples available to the pathologist.

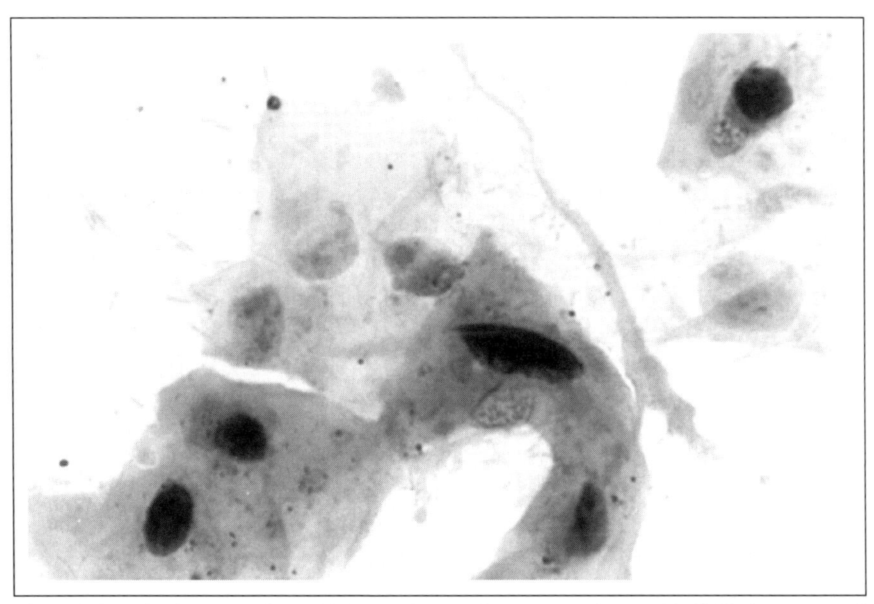

Figure 2. Imprint from a patient with known low grade squamous intraepithelial lesion processed as above.

The use of direct incorporation of labeled nucleotides during the polymerization process, essential to indirect *in situ* PCR, is not to be recommended since fragmented DNA, caused by apoptosis or other forms of cell death, may serve as primer and induce an extension reaction leading to false-positive results. Endogenous mispriming or repair artifacts are further pathways for this kind of reaction (12,17, 18). The reason we did not encounter this problem in SiHa cells is probably due to the fact that these cells no longer induce apoptosis in culture. We, therefore, advise that ISH with specific, labeled probes should replace the direct incorporation method for the detection of the PCR-amplicons. The positive aspect about direct incorporation, however, is that we could actually prove that PCR *inside the nucleus* takes place!

The technical improvement of our assay was necessary to obtain reproducible and reliable results. When testing a conventional PCR instrument for PISH, our measurements showed that there was a substantial loss of temperature. We similarly noticed significant variations in temperature distribution, depending on how the slide was placed on the perforated thermo-block. By employing a thermal cycler with a solid thermo-block for glass slides, we were able to eliminate this problem completely. We, therefore, consider traditional thermal cyclers to be unsuitable for PISH. Our way of preventing evaporation, by using a special setup of a wax circle surrounding the cytospins or tissue sections and mineral oil with two consecutive coverslips, is considerably easier and faster compared to the rather tedious nail polish procedure and lifting up the coverslip for "hot start" as previously reported (13–16).

PISH has been developed primarily to improve sensitivity of conventional ISH for viral detection, leading to a better understanding of the behavior of these microorganisms.

REFERENCES

1. **Bagasra, O., S.P. Hauptman, H.W. Lischner, M. Sachs and R.J. Pomerantz.** 1992. Detection of human immunodeficiency virus type 1 in mononuclear cells by *in situ* polymerase chain reaction. N. Engl. J. Med. *326*:1385-1391.
2. **Bagasra, O. and R.J. Pomerantz.** 1993. Human immunodeficiency virus type 1 provirus is demonstrated in peripheral blood monocytes in vivo: A study utilizing an *in situ* polymerase chain reaction. AIDS Res. Hum. Retroviruses *9*:69-76.
3. **Bagasra, O., T. Seshamma and R.J. Pomerantz.** 1993. Polymerase chain reaction *in situ*: Intracellular amplification and detection of HIV-1 proviral DNA and other specific genes. J. Immunol. Methods *158*:131-145.
4. **Bauer, H.M., C.E. Greer and M. Manos.** 1992. Determination of genital human papillomavirus infection by consensus PCR amplification, p. 131-152. *In* C.S. Herrington and J.O'D. McGee (Eds.), Diagnostic Molecular Pathology, Vol. II. Oxford University Press, Oxford.
5. **Chiu, K.P., S.H. Cohen, D.W. Morris and G.W. Jordan.** 1992. Intracellular amplification of proviral DNA in tissue sections using the polymerase chain reaction. J. Histochem. Cytochem. *40*:333-341.
6. **Gosden, J. and D. Hanratty.** 1993. PCR *in situ*: A rapid alternative to *in situ* hybridization for mapping short, low copy number sequences without isotopes. BioTechniques *15*:78-80.
7. **Haase, T., E.F. Retzel and K.A. Staskus.** 1990. Amplification and detection of lentiviral DNA inside cells. Proc. Natl. Acad. Sci. USA *87*:4871-4875.
8. **Hacker, G.W., L. Grimelius, G. Danscher, G. Bernatzky, W. Muss, H. Adam and J. Thurner.** 1988. Silver acetate autometallography: An alternative enhancement technique for immunogold

silver-staining (IGSS) and silver amplification of gold, silver, mercury and zinc in tissues. J. Histotechnol. *11*:213-221.

9. **Hacker, G.W., A.-H. Graf, C. Hauser-Kronberger, G. Wirnsberger, A. Schiechl, G. Bernatzky, U. Sonnleitner-Wittauer, H. Su, H. Adam, J. Thurner, G. Danscher and L. Grimelius.** 1993. Application of silver acetate autometallography and gold-silver staining methods for *in situ* DNA hybridization. Chin. Med. J. *106*:83-92.
10. **Komminoth, P. and A.A. Long.** 1993. In situ polymerase chain reaction. An overview of methods, applications and limitations of a new molecular technique. Virchows Arch. [B] *64*:67-73.
11. **Komminoth, P., A.A. Long, R. Ray and H.J. Wolfe.** 1992. In situ polymerase chain reaction detection of viral DNA, single copy genes and gene rearrangements in cell suspensions and cytospins. Diagn. Mol. Pathol. *1*:85-97.
12. **Long, A.A., P. Komminoth, E. Lee and H.J. Wolfe.** 1993. Comparison of indirect and direct in-situ polymerase chain reaction in cell preparations and tissue sections. Detection of viral DNA, gene rearrangements and chromosomal translocations. Histochemistry *99*:151-162.
13. **Nuovo, G.J., F. Gallery, P. MacConnell, J. Becker and W. Bloch.** 1991. Detection of human papillomavirus DNA in formalin-fixed tissues by *in situ* hybridization after amplification by polymerase chain reaction amplification. Am. J. Pathol. *139*:1239-1244.
14. **Nuovo, G.J., F. Gallery, P. MacConnell, J. Becker and W. Bloch.** 1991. An improved technique for the *in situ* detection of DNA after polymerase chain reaction amplification. Am. J. Pathol. *139*:1239-1244.
15. **Nuovo, G.J., F. Gallery and P. MacConnell.** 1992. Detection of amplified HPV 6 and 11 DNA in vulvar lesions by hot start PCR *in situ* hybridization. Mod. Pathol. *5*:444-448.
16. **Nuovo, G.J., M. Margiotta, P. MacConnell and J. Becker.** 1992. Rapid *in situ* detection of PCR-amplified HIV-1 DNA. Diagn. Mol. Pathol. *1*:98-102.
17. **Sällström, J.F., I. Zehbe, M. Alemi and E. Wilander.** 1993. Pitfalls of *in situ* PCR using direct incorporation of labeled nucleotides. Anticancer Res. *13*:1153-1154.
18. **Sällström, J.F., M. Alemi, H. Spets and I. Zehbe.** 1994. Nonspecific amplification in *in situ* PCR by direct incorporation of reporter molecules. Cell Vision *1*:243-251.
19. **Schneider, A., G. Meinhardt, R. Kirchmayr and V. Scheider.** 1991. Prevalence of human papillomavirus genomes in tissues from the lower genital tract as detected by molecular *in situ* hybridization. Int. J. Gynecol. Pathol. *10*:1-7.
20. **Spann, W., K. Pachmann, H. Zabienska, A. Pielmeier and B. Emmerich.** 1991. *In situ* amplification of single copy gene segments in individual cells by the polymerase chain reaction. Infection *19*:242-244.
21. **Staskus, K.A., L. Couch, P. Bitterman, E.F. Retzel, M. Zupanic, J. List and A.T. Haase.** 1991. In situ amplification of visna virus DNA in tissue sections reveals a reservoir of latently infected cells. Microb. Pathog. *11*:67-76.
22. **Ting, Y. and M.M. Manos.** 1990. Detection and typing of genital human papillomaviruses. *In* M.A. Innis et al. (Eds.), PCR Protocols. A Guide to Methods and Applications. Academic Press, San Diego.
23. **Zehbe, I., E. Rylander, A. Strand and E. Wilander.** 1992. *In situ* hybridization for the detection of human papillomavirus (HPV) in gynaecological biopsies. A study of two commercial kits. Anticancer Res. *12*:1383-1388.
24. **Zehbe, I., E. Rylander, A. Strand and E. Wilander.** 1993. Use of Probemix and Omniprobe biotinylated cDNA probes for detecting HPV infection in biopsy specimens from the genital tract. J. Clin. Pathol. *46*:437-440.
25. **Zehbe, I., G.W. Hacker, J. Sällström, E. Rylander and E. Wilander.** 1992. Detection of single HPV copies in SiHa cells by *in situ* polymerase chain reaction (*in situ* PCR) combined with immunoperoxidase and immunogold silver-staining (IGSS) techniques. Anticancer Res. *12*:2165-2168.
26. **Zehbe, I., J. Sällström, G.W. Hacker, E. Rylander, A. Strand, A.-H. Graf and E. Wilander.** Polymerase chain reaction (PCR) *in situ* hybridization: Detection of human papillomavirus (HPV) DNA in SiHa cell monolayers. *In* J. Gu and G.W. Hacker (Eds.), Modern Analytical Methods in Histochemistry (In press).

Address correspondence to Ingeborg Zehbe, Department of Pathology, University Hospital, S-751 85 Uppsala, Sweden.

Conventional PCR, *In Situ* PCR and Reverse Transcription *In Situ* PCR for HIV Detection

Eduardo A. Zevallos[1,2], Enzo Bard[2], Virginia M. Anderson[2], Tak-Shun Choi[1,2] and Jiang Gu[1]

[1]Deborah Research Institute, Browns Mills, NJ and [2]SUNY Health Science Center at Brooklyn, NY, USA

SUMMARY

Recent developments in the detecting technology of in situ *PCR, reverse transcription* in situ *PCR, conventional PCR and* in situ *hybridization have advanced our understanding of the pathogenesis of AIDS. With those techniques, minute quantities of HIV sequences were localized to specific cell types in various organs at different stages of disease development. New discoveries using those technologies in AIDS research are unveiling the whole picture of HIV infection. This chapter discusses the background of HIV,* in situ *PCR and the related techniques. It provides theoretical and practical considerations and detailed step-by-step protocols for HIV detection based on our own experiences. It is intended to give the reader a comprehensive review and a practical guide for AIDS research using these newly developed morphological technologies.*

INTRODUCTION

Just over a decade ago, the acquired immunodeficiency syndrome (AIDS) was first diagnosed in homosexual males in the United States (6) and has since become an escalating heterosexual epidemic affecting the global community. Groups at risk for human immunodeficiency virus type 1 (HIV-1) infection include intravenous drug addicts, sex partners of intravenous drug addicts, offspring of HIV-infected mothers, homosexual men and recipients of blood products. An estimated 1 to 1.5 million people are infected by HIV in the US alone (16,31). By 1995, approximately ten million people will be infected worldwide with the virus. Hence, understanding the pathogenesis of AIDS in different organs is extremely urgent to curtail this deadly disease. Preservation of tissue morphology is critical in defining the cell populations harboring HIV-1 and their roles in AIDS pathogenesis.

The main target of HIV-1 is the immune system (13). Its best known immunopathogenesis is the depletion of T4 lymphocytes that express the CD4 molecules (8,9,22,30,32). The loss of these cells severely handicaps the cellular and humoral immunological responses. Cells expressing CD4 molecules other than lymphocytes are also permissive to HIV-1 infection (10,12,15,17,35,39,40). With the new evidence, HIV-1 infection must be considered a multisystemic pathological process that affects almost every organ. Some pathological lesions such as encephalopathy, cardiomyopathy, enteropathy and nephropathy have been suspected of being the direct effects of HIV-1 infection. As the spectrum of HIV-1 infection in different organs has not been fully defined, any pathological classification of HIV-1 lesions must be considered preliminary. More importantly, a sensitive and reliable method of detecting minute quantities of HIV viral sequences morphologically is needed.

Conventional Detection Methods of HIV Infection

Immunohistochemistry and *in situ* hybridization are inconclusive for the detection of HIV-1 at the cellular level. This is probably attributable to the relatively low sensitivity of these methods to detect HIV viral antigens or sequences. Immunohistochemistry using different types of anti-HIV antibodies has produced variable results. They are affected by the types of antibodies, detection systems, tissue fixation, the amount of antigens per cell and the availability of expressed HIV-1 epitopes. It is often necessary to use freshly frozen sections for preservation of HIV-1 antigenicity. Usually, only a small proportion of cells harboring HIV-1 have detectable viral mRNA (29) and, consequently, they may not express enough of the related proteins. *In situ* hybridization requires about twenty copies of target sequences per cell for detection (23,36). The number of HIV genomes per cell is usually below this detection threshold (1,27).

Based on our experience using a highly sensitive *in situ* polymerase chain reaction (ISPCR) on AIDS autopsy, variations in the amount of HIV-1 burden in various organs of different patients are apparent. For instance, pediatric AIDS cases with lymphoproliferative syndrome display viral positivity in most lymphoid cells. On the other hand, the positivity in lymphoid cells is sparse in AIDS cases with lymphoid depletion (44). When using low sensitivity methods such as immunohistochemistry and *in situ* hybridization, even under optimal conditions, there is a significant number of false negative results.

Roles of PCR in Detecting HIV Sequence

PCR is a highly sensitive method capable of detecting minute amounts of target nucleic acid sequences, as is the case in HIV-1 infection. PCR is a cyclic process by which an exponential amplification of a target sequence is obtained. After 20 cycles, a single copy of DNA can result in a millionfold of amplicons (34). With this power of amplification, PCR has successfully detected HIV-1 in peripheral blood of adult and pediatric infected patients (14,20,21,28,33). However, since conventional PCR is performed on extracted DNA or RNA from homogenized tissue, it is not able to distinguish the population of cells harboring the target HIV sequences. ISPCR, a newly developed method, is capable of amplifying the target DNA or RNA sequences on tissue sections or cells *in situ*. It is more sensitive than *in situ* hybridization. PCR *in situ* hybridization, combined with flow cytometry on peripheral mononuclear cells from HIV-infected patients, has been able to detect HIV-1 DNA and RNA in individual cells (29). The expression of HIV-1 RNA as an indicator of active infection is important in assessing the clinical outcome of the patients.

ISPCR does not have the sensitivity and efficiency of conventional PCR when expressed as the absolute number of amplified target sequences per reaction. However, comparison of both methods should be based on the number of amplicons per cell. ISPCR is influenced by the same range of factors that affects the outcome of conventional PCR. These include temperature, number of cycles, concentration of individual reagents and the inherent plateau effect of the reaction. They are more difficult to manage in ISPCR than in conventional PCR.

ISPCR

Fixation

For ISPCR, tissue samples should be fixed as fresh as possible. Ten percent buffered formalin is an adequate fixative for ISPCR and is available in most morphology laboratories. Formalin and paraformaldehyde have been the most common fixatives used to preserve samples for ISPCR. Their cross-linking property has been suggested as desirable for retention of amplicons at the site of origin (25). Prolonged fixation with cross-linking fixatives, however, is not recommended because it may render the target sequence unamplifiable. Excessive cross-linking between the strands of nucleic acids with proteins produces significant molecular alterations. Based on our experience, the quantity and quality of DNA or RNA extracted from formalin over-fixed tissues are poor. Instead, coagulative fixatives such as absolute ethanol preserve the integrity of nucleic acids for long periods of time. We have successfully extracted DNA from tissues preserved in ethanol for two years. On the other hand, it has been suggested that ethanol fixation is not ideal for ISPCR, as the lack of cross-linking leads to excessive elution of amplicons (25). It appears that the proper amount of cross-linking of nucleic acids is critical for the outcome of ISPCR. We recommend 4 to 6 h fixation in 10% buffered formalin. If a longer period of fixation is unavoidable, i.e., if it is not possible to process a tissue sample the same day, it is advisable to initially fix the specimen in absolute ethanol and then in 10% buffered formalin before embedding.

Enzyme Digestion

For *in situ* amplification, the target sequence under investigation is embedded in the tissue sections or cells, and unmasking is required. Depending on the nature of the specimen, the intensity of fixation and the type of fixative used, protease digestions are needed. The most common digestion method is pretreatment with proteinase K at different concentrations, durations and temperatures. Over-digestion may produce morphological distortions to the point that interpretation of results becomes impossible. The presence of membrane "blobs" on lymphocytes has been indicated as a sign of appropriate protease digestion (2).

Elution of Amplicons and False Positivity

During ISPCR, a significant number of amplicons elute from the point of origin. This undesirable phenomenon is caused by the physical characteristics of short sequences in an aqueous phase at high temperatures that has increased molecular motion. In the case of tissue sections with interrupted cell membranes, the proportion of eluted amplicons could be very significant. Increasing the thickness of tissue sections may palliate this problem, but the possibility that eluted amplicons may bind to other sites is increased. This false positivity may show up at a different cellular location or in another cell type and create misinterpretation of the results. The elution of amplicons is significantly reduced when ISPCR is performed in intact cells, as the amplicons are contained by preserved cell membranes. It is possible that even when using intact cells, there is diffusion of amplicons to the aqueous phase, which becomes the source for false positivities.

SIGNIFICANCE OF "LONG PRODUCTS" IN ISPCR

Recently, it has been suggested that PCR "long products" are the most important amplified sequences for the ISPCR method (11). These "long products", complementary to the native strands, are nucleotide sequences extended beyond those encased by the primers and are synthesized at an arithmetical rate. Elution is less likely to occur for the "long products" than the short amplicons. Furthermore, complementary "long products" may anneal among themselves and, with complementary amplicons in abundance, avert further elution and increase the sensitivity of the method. For instance, starting the PCR with one target double-stranded DNA sequence and assuming 100% efficiency, after 30 cycles a billionfold amplification of amplicons is expected. For the same PCR conditions, 60 "long products" will be generated, rendering positivity in the subsequent *in situ* hybridization step. Although the "long products" are disregarded in the conventional PCR because of their insignificant yield and heterogeneous dimensions, they may be extremely meaningful for the sensitivity of ISPCR. If this is indeed the case, performing ISPCR at a higher number of cycles may be advantageous. Moreover, using a single primer will create only the "long products" without the abundant "short amplicons" to compete for reagents. However, one type of strand will not form DNA networks that can anchor the products *in situ* without an equal number of the complementary strands.

POST-AMPLIFICATION WASHINGS

Stringent washings after the *in situ* amplifications are recommended to remove free amplicons in the liquid-phase overlying the sample (42,43). Evaporation or ethanol washings after the *in situ* amplification should be avoided. Ethanol produces precipitation of DNA including amplicons and primers and leads to background or false positivity. Nevertheless, some authors reported good results with ethanol washings (3,41). We have observed that excessive stringency removes products of amplification attached to the cellular structures. This is a very critical step for *in situ* PCR performed on tissue sections. In addition, the method of detection may substantially affect the results of *in situ* PCR. Depending on their stringency, washings may provoke significant elution of bound amplification products. We recommend washing with stringent standard saline citrate (SSC) solutions and phosphate-buffered saline (PBS), and avoid washings with distilled water after the amplification. If counterstain is required, it is recommended that the dye be dissolved in SSC or PBS.

DETECTION OF *IN SITU* AMPLIFIED SEQUENCES

It is the consensus of most experts that the detection of *in situ* amplified sequences should be revealed by *in situ* hybridization using an oligonucleotide probe that increases the specificity of the method. The probe must be complementary to an inner sequence of the amplified target sequence. Double-stranded probes are suitable because the amplification products are double stranded. Both the probes and the amplified signals require heat denaturation before hybridization. Direct incorporation of labeled nucleotides during ISPCR on tissue sections is discouraged by most investigators because it tends to produce mislabeling. This undesirable consequence may be related to a repairing process occurring during *in situ* PCR. The tissue processing and sectioning may damage the indigenous genomic DNAs, which in turn may behave as pseudo-primers. Directly labeled ISPCR has been performed on cytological specimens with good results. This is probably because intact cells have undamaged DNA. However, we still recommend using a specific probe to detect the amplified sequences. Directly labeled *in situ* PCR has been used to verify the *in situ* amplifiability of DNA, but the positivity may be caused by DNA repair rather than specific amplification. For instance, it has been observed that apoptotic cells produced intense positivity, which might be caused by damaged DNA. Conventional PCR or *in situ* PCR of indigenous DNA or commonly expressed RNA are useful to verify the amplifiability of samples and to optimize the procedure as positive controls (4,5,38,37). Reverse transcribed *in situ* PCR (RT-ISPCR) should be performed on DNase-pretreated samples that have their entire cellular DNA destroyed to prevent pseudo positivity. Since RNA is single stranded and is not reparable by the enzymes used for RT-ISPCR, no mislabeling should result. We performed directly labeled RT-ISPCR to verify the presence and amplifiability of mRNA with primers for commonly expressed genes (i.e., retinoblastoma) with good results (45,46). However, we do not recommend the direct label method in HIV RT-ISPCR for conserved gene sequences, as pseudo-primer DNA reparative effect may occur even in DNase pretreated samples. The intensity of fixation and the extent of DNA cross-linking may preclude complete degradation of DNA by DNase, and the pseudo-primer reparative effect may not be completely avoided. The interpretation of the results of *in situ* PCR should be made by someone with histopathological training and a molecular biology background.

Controls in ISPCR

HIV ISPCR experiments should be performed in parallel with conventional PCR using the same sample and the same primers. In addition to HIV-1 positive and negative controls, internal controls must be performed in each ISPCR assay. For ISPCR, they include DNase pretreatment and assays omitting primers, *Taq* DNA polymerase or probes. For RT-ISPCR, they include RNase pretreatment and assays omitting random hexamers, "downstream" primer or reverse transcriptase. Other internal controls such as omitting buffer, $MgCl_2$, deoxyribonucleoside triphosphates (dNTPs) or other ingredients may also be performed. Our HIV ISPCR technique has been optimized with multiple negative, positive and internal controls, which are imperative for obtaining reliable results (46,48). Positive and negative HIV-1 controls must be selected before starting the ISPCR experiments. We regularly obtain our HIV-1 tissue controls from fresh autopsies with known HIV status. A battery of frozen specimens, absolute ethanol-fixed tissues and formalin-fixed paraffin-embedded blocks are maintained as controls. Specificity of the ISPCR can also be verified using probes that are noncomplementary to the amplified target sequence. Comparing results of ISPCR with those from *in situ* hybridization without amplification is recommended.

Modifications of ISPCR Technique

Modifications have been made in our *in situ* PCR technique as compared with other protocols (1,3,7,18,19,41). These include pre-PCR prehybridization, "cold-hot-start" technique and post-PCR heating. We believe that the prehybridization and post-heating modifications enhance the adhesiveness and retention of amplified products in the tissue sections or cells. The prehybridization solution consists of dextran sulfate, SSC, EDTA and salmon sperm DNA. The solution is made without formamide because formamide may either affect the thermodynamics of PCR or inhibit the *Taq* DNA polymerase. We speculate that the dextran in the prehybridization solution may have a "glue" effect that facilitates the retention of amplification products at the original site.

A pre-cooled PCR "master mix" is added to preheated slides to perform a "hot start" at 70°C. At such a high temperature, it is unlikely that the primers will anneal to nonspecific sequences. To avoid mispriming, the sections are maintained in the thermocycler at this temperature and the heat cycles start at the denaturing point. By using this "cold-hot-start" technique, we circumvent the lifting of the glass coverslips for adding a second portion of the PCR mixture, which is performed in other protocols. Once the thermocycling is completed, stringent washings are immediately carried out to remove free elements contained in the liquid phase, which otherwise may create background or false positivity. Post-PCR baking is performed in an attempt to enhance the attachment of the amplified PCR products to the tissue elements, thereby further retaining the amplified products. The same beneficial effect of heating has also been reported by other investigators (3).

The recently developed slide thermocyclers (e.g., PTC-100™; MJ Research, Watertown, MA, USA) allow appraisal of many samples and different controls simultaneously. When a tube thermocycler (Perkin-Elmer, Norwalk, CT, USA) is adapted for ISPCR, it is necessary to cover the block with aluminum foil, add mineral oil in the holes and press the slides against the block to optimize temperature transmission. Even so, homogeneous temperature on the slides is not guaranteed. The block of a tube thermocycler usually accommodates only a few slides and thus limits the capacity of ISPCR experiments.

CONCLUSIONS

The key points of *in situ* PCR for HIV-1 detection are preservation of cell and tissue morphology, retention of amplicons at the site of origin and specificity of the amplified PCR products. Our experience in HIV ISPCR is primarily based on the study of placentas of HIV-infected mothers (42). We have detected HIV-1 in other types of tissues and cells using the same *in situ* PCR technique. These included peripheral mononuclear cells, bronchial-alveolar lavage cells, lymph nodes, spleen, cervix and lung (32,44,48). *In situ* PCR is currently considered the morphological method of choice for detection of HIV-1 at the cellular level (1,11,24,26,30,42,43). HIV ISPCR is important for elucidating the infectious status and the population of cells harboring the virus. HIV-infected cells usually display nuclear positivity, which indicates that the viral genome has been incorporated into the host genome. Less frequently we have observed cytoplasmic positivity, which may suggest an active state for the infection (3,42). In our experiments with tissues and cells from HIV-infected subjects, we obtained comparable results from ISPCR and conventional PCR for detection of HIV-1 sequences. The specificities of the two methods are similar. The sensitivity of HIV ISPCR, as mentioned above, has not been clearly defined, as quantitative *in situ* amplification has not been attempted on tissue sections. In our experience on placentas of HIV-infected mothers, nested PCR employing more than one pair of primers, which was supposed to increase both sensitivity and specificity, did not produce additional information. In our hands, HIV ISPCR and conventional PCR running in parallel are sensitive enough for low copy number HIV detection and accurate diagnosis.

RECOMMENDED PROTOCOLS

Infectivity of Embedded Tissue and PCR Products

DNA and RNA extractions can be performed efficiently from fixed and paraffin-embedded tissue (36,38). The risk of transmission to investigators is insignificant when using this material. RNA can also be efficiently retrieved from protease-digested tissue samples without using elaborate procedures (5). The commercial kit for HIV PCR (GeneAmplimer® HIV-1 Control Reagents; Perkin-Elmer, Norwalk, CT, USA) does not contain infectious material. The products of amplification are also noninfectious.

Contamination and Optimization

PCR experiments should be handled by trained investigators to minimize technical errors and contamination of reagents. To avoid contamination with amplified PCR products, it is highly advisable that tissue extractions and PCR mixtures be handled in a room separate from the thermocycler. Multiple experiments using different concentrations of reagents are necessary to achieve the optimal PCR efficiency.

DNA Extraction

The following procedures for DNA and RNA extractions can be applied to fresh, formalin- or ethanol-fixed and paraffin-embedded tissue. Minute amounts of tissue samples are required.

From fresh or fixed tissue, 3–4 mm^3 sampling is enough. From paraffin-embedded tissue, 2–4 5-μm-thick sections are adequate.

Method A. Phenol/Chloroform Extraction of DNA

1. Place tissue sample into a 0.5-mL microcentrifuge tube. If the sample is paraffin-embedded material, deparaffinize as follows. Add 400 μL of xylene for 10 min at room temperature. Spin in a microcentrifuge at 12 000× *g* for 5 min to pelletize the tissue. If the xylene is cloudy, remove it, add xylene and repeat the centrifugation. Decant the supernatant, add 400 μL of 100% ethanol and centrifuge at 12 000× *g* for 5 min. Repeat this step at least twice or until residual ethanol is translucent. Desiccate the sample in a vacuum chamber for 1 h.
2. Add 100 μL of DNA extraction buffer A to the tissue pellet. Disrupt the tissue with the tip of pipet and incubate at 54°C overnight. The DNA extraction buffer A consists of a solution containing 100 mM NaCl, 10 mM Tris-HCl, pH 8.0, 25 mM EDTA, pH 8.0, 0.5% sodium dodecyl sulfate (SDS) and 3 mg/mL proteinase K.
   Note: Fresh tissue may require less incubation time, but a minimum of 2 h is advised. Formalin-fixed tissue may require prolonged incubation depending on the intensity of previous fixation. Sometimes 2 to 3 days are required, but more than 5 days may damage the DNA.
3. Heat the samples in a block at 95°C for 7 min to inactivate the proteinase K.
4. Centrifuge the digested sample at 12 000× *g* for 5 min. Transfer the supernatant to a 0.5-mL microcentrifuge tube.
5. Add equal volume of phenol and vortex mix for 5 min. Centrifuge at 12 000× *g* for 5 min and transfer the supernatant to a 0.5-mL microcentrifuge tube.
6. Add equal volume of chloroform and vortex mix for 3 min. Centrifuge at 12 000× *g* for 5 min and transfer the supernatant to a 0.5-mL microcentrifuge tube.
7. Add double volume of absolute ethanol. If precipitation of DNA is visible, recover it by stirring and transfer to a microcentrifuge tube. If no precipitation is visible, refrigerate the tube with absolute ethanol at 4°C overnight, then centrifuge at 12 000× *g* for 5 min and decant the ethanol. The pellet contains the DNA.
8. Resuspend the pellet in 50 to 100 μL of TE buffer at 4°C overnight. TE buffer consists of 10 mM Tris-HCl, 1 mM EDTA, pH 7.5.
9. Store at 4°C.

Method B. Non-Phenolic/Chloroform DNA Extraction (18) (modified)

1. Same as in method A.
2. Add 100 μL of DNA extraction buffer B to the microcentrifuge tube and vigor-

ously disrupt the pellet with the tip of the pipet. Incubate at 54°C overnight. Different digestion periods may be necessary according to the nature and fixation of the samples. The DNA extraction buffer B is a solution of 100 mM Tris-HCl, pH 8.0, 1 mM EDTA, pH 8.0, and 0.3 mg/mL proteinase K.
3. Heat the samples in a block at 95°C for 7 min to inactivate the proteinase K.
4. Centrifuge at 12 000× *g* for 5 min. Transfer the supernatant containing the DNA to a microcentrifuge tube.
5. Store at 4°C.

DNA ASSESSMENT

Spectrophotometry

Make a dilution (1/50, 1/100, etc.) of the DNA samples in double-distilled H_2O (ddH_2O). Measure the optical densities at 260 nm. Calculate the concentration and the total amount of DNA. Calculate the optical densities (OD) at 280 nm to assess the protein contaminant. Calculate the OD ratios to OD260/OD280. If this ratio is much less than 2.0, consider that a significant amount of protein is still present and the extraction technique needs optimization.

Electrophoresis

Perform electrophoresis on 0.7% LE-agarose in 1× TAE buffer containing ethidium bromide. Each well of the gel is filled with 10 µL of the DNA. A molecular weight marker, i.e., Molecular Weight Marker II (Boehringer Mannheim, Indianapolis, IN, USA), is simultaneously electrophoresed. Once completed, the gel is analyzed under UV light, and Polaroid exposures are taken as necessary. Usually the DNA observed as a smear in the area nearest to the cathode are the largest molecular-weight fragments. For PCR purposes, the larger the fragments, the better the expected amplifiability. If DNA targets for PCR amplification are in the range of 100 to 400 bp, yielded DNA of more than 500 bp may be adequate.

PCR of Indigenous Genomic Sequences

We regularly perform conventional PCR using Beta-globulin primers, PCO_4 and GH_{20} (Perkin-Elmer), on each DNA sample to confirm its amplifiability. These primers amplify a target sequence of 268 bp (38). If a sample does not produce this band, it is excluded from further analysis.

RNA EXTRACTION (38) (modified)

1. Same as in Method A for DNA extraction.
2. Add 100 µL of RNA extraction buffer, which consist of 1.1 M guanidinium isothiocyanate, 0.56% Sarkosyl, 22 mM Tris-HCl and 0.3 mg/mL proteinase K. Disrupt the tissue with the tip of the pipet.
3. Incubate at 45°C overnight and then inactivate proteinase K at 95°C for 7 min.
4. Extract with an equal volume of a solution of 70% phenol (equilibrated in ddH_2O) and 30% chloroform; gently mix for 1 min.

5. Centrifuge at 12 000× *g* for 15 min. Transfer the supernatant to a 0.5-mL microcentrifuge tube.
6. Add an equal volume of isopropanol and precipitate at -20°C overnight.
7. Centrifuge at 12 000× *g* for 15 min and decant the supernatant. The pellet contains the RNA.
8. Wash the pellet with 70% ethanol and air-dry.
9. Dissolve the RNA sample in 25 µL of TE buffer.

RNA ASSESSMENT

The amplifiability of extracted RNA can be appraised by performing RT-PCR. Primers for retinoblastoma mRNA are used, as reported elsewhere (38).

HIV CONVENTIONAL PCR

Primers

The amplification of HIV-1 sequences is carried out by using two commercial kits: GeneAmp® PCR Core Reagents and GeneAmplimer HIV-1 Control Reagents (Perkin-Elmer). They contain all the elements for HIV PCR. The SK38 and SK39 primers that amplify a 115-bp conserved sequence in the *gag* region of HIV-1 are supplied in the kits. We have successfully used other pairs of HIV primers, SK102 and SK431, which amplify a 145-bp conserved *gag* sequence.

Hot Start

The "hot start" as recommended by the manufacturer (AmpliWax™; Perkin-Elmer) is performed in our experiments. The principle of this technique is to start the PCR with all the elements including the DNA sample at high temperature. Using an intermediate wax layer permits withholding one of the main reagents (primers, DNA or *Taq* DNA polymerase) on the "upper layer" in the same PCR tube until the temperature of the system is about 75°C. At this temperature, the wax melts and the "upper layer" reagents mix with the "lower" reagents. The total mixture volumes of the wax gems are commercially available in 100 and 50 µL. (If lower volumes are used, we recommend heating the "lower" reagent mixture overlaid with mineral oil in the thermocycler, and then adding the "upper" reagent.) At small volumes, the wax may be less efficient in avoiding evaporation, and with a relatively large mass of wax, the mixing is impaired and pipetting is difficult.

Controls

Before preparing the mixtures, one should make all the calculations. The number of experimental samples is added to the number of controls. Every PCR assay must include positive and negative controls as a minimum requirement. The kit for HIV-1 amplification includes a plasmid DNA with complete HIV-1 genome. The negative control consists of using non-HIV-infected human placental DNA and assaying less than twenty samples at a time to avoid technical errors and cross-contamination. Beginners must be supervised by trained technicians or investigators.

The protocol for HIV PCR is as follows:

1. The "lower" reagent mix consists of a final concentrated solution of 1.25× PCR Buffer II (Perkin-Elmer), 2.5 mM $MgCl_2$, 200 μM each dNTP, 0.5 μM of each *gag* primer. For a total volume of 12.5 μL per reaction, we used the kit components as follows: 1.25 μL 10× PCR Buffer II, 5.0 μL 25 mM $MgCl_2$, 1 μL each 10 mM dNTP, 1 μL each 25 μM *gag* primer, and 0.25 μL ddH_2O.
2. The "upper" reagent mixture consists of a final concentrated solution of 1.25× PCR Buffer II, *Taq* DNA polymerase 1.25 units/reaction. (For a total volume of 32.5 μL, we used 27.25 μL ddH_2O, 5.0 μL 10× PCR Buffer II and 0.25 μL 5 U/μL *Taq* DNA polymerase.) The total volume of "upper" mix, "lower" mix and DNA aliquot is 50 μL. The volume can be decreased and the proportion of components in each mix can be changed, but the final concentration of each reagent must be maintained as recommended by the manufacturers.
3. Place the "lower" reagent mixture in the bottom of the PCR tubes; add one unit of wax and place the tubes in a hot block at 70° to 80°C until the wax melts (in approximately 1 min or less). Remove the tubes from the block and cool at room temperature or at 4°C until the wax forms a solid overlying layer.
4. Add the "upper" reagent mixture on the wax layer. Both mixtures are separated by the intermediate wax without intermixing.
5. The DNA samples and controls are added in the "upper" reagent mixture. For a total volume mixture of 50 μL, we use 5 μL of experimental sample containing approximately 0.5 μg of DNA, and 5 μL of HIV-1-positive and HIV-1-negative control DNA.
6. Transfer the PCR tubes to a thermocycler. We used these temperature parameters: 95°C for 2 min (initial denaturing), 35 cycles of 95°C for 45 s (denaturing), 55°C for 60 s (annealing) and 72°C for 45 s (extension), followed by 72°C for 3 min (final extension). Finally, temperature is decreased to 4°C (soaking).
7. Ten microliters of each PCR product and 3 μL of appropriate molecular weight marker (Molecular Weight Marker V; Boehringer Mannheim, Indianapolis, IN, USA) are simultaneously electrophoresed in NuSieve®-LE agarose 3:1 in TAE buffer containing ethidium bromide. The gels are examined under UV light to observe the expected DNA bands, and Polaroid photos are taken as required.

RT-PCR

The RT-PCR is carried out by using commercial kits according to the manufacturer's instructions. We have experience with two kits (GeneAmp RNA PCR Kit and Thermostable r*Tth* Reverse Transcriptase RNA PCR Kit; Perkin-Elmer), which are equally effective. However, we prefer the r*Tth* because this enzyme performs the reverse transcription and the PCR in the same assay, and the reversed transcription operates at 70°C with the "downstream" primer without using random hexamers or oligo (dT).

Method A. Reverse Transcriptase-Driven RT-PCR

1. Prepare a RT mixture as indicated per reaction in a PCR tube and overlay it with one drop of mineral oil.

	Volume	Final Concentration
$MgCl_2$	2 µL	5 mM
10× PCR Buffer II	1 µL	1×
ddH_2O	0.5 µL	-
dATP	1 µL	1 mM
dCTP	1 µL	1 mM
dGTP	1 µL	1 mM
dTTP	1 µL	1 mM
RNase inhibitor	0.5 µL	1 U/µL
Reverse transcriptase	0.5 µL	2.5 U/µL
Random hexamers	0.5 µL	2.5 µM
or		
Oligo (dT)	0.5 µL	2.5 µM
or		
"Downstream" primer	0.5 µL	0.75 µM
Positive control RNA	1 µL	10(4) copies
or		
Experimental sample	1 µL	=<1 µg RNA
Total volume	10 µL	

2. If random hexamers are used, incubate the tubes at room temperature for 10 min. This incubation is not required if oligo (dT) or "downstream" primers are used to assemble the cDNA.
3. Transfer the tubes to a thermocycler programmed as follows: 1 cycle of 42°C 15 min, 99°C 5 min and 5°C 5 min.
4. Add 39 µL of PCR mixture to each tube. This mixture must be prepared in advance as indicated.

	Volume	Final Concentration
$MgCl_2$	2 µL	2 mM
10× PCR Buffer II	4 µL	1×
ddH_2O	32.75 µL	-
Taq DNA polymerase	0.25 µL	1.25 U/50 µL
Total volume	39 µL	

5. Dispense the 0.5 µL of each primer into each tube. If "downstream" primer is used for the reverse transcription, do not add it again and substitute with 0.5 µL of ddH_2O.
6. Perform the PCR in a thermocycler with the next parameters: 2 min at 95°C for 1 cycle (denaturing). Thirty-five cycles of 95°C 45 s, 55°C 1 min, 72°C 45 s. Finally, 4°C (soaking).

7. Store the amplified samples at -20°C until further analysis.
8. Run electrophoresis as described in the PCR protocol. Southern blot analysis may be necessary if bands are not visualized in the gel.

Method B. r*Tth* DNA Polymerase-Driven RT-PCR

1. Prepare the RT mixture per reaction as follows:

	Volume	Final Concentration
ddH_2O	5.2 µL	-
10× RT Buffer	1 µL	1×
$MnCl_2$	1 µL	1.0 mM
dNTPs each	0.2 µL	200 µM
r*Tth* DNA polymerase	1 µL	2.5 U/10 µL
"Downstream" primer	0.5 µL	0.75 µM
Experimental sample	0.5 µL	=<125 ng RNA
Total volume	10 µL	

2. Overlay the tube with one drop of mineral oil.
3. Incubate the tubes in a thermocycler at 70°C for 15 min. Stop the reaction at 4°C.
4. Add 40 µL of PCR mixture to each tube. This mixture must already be prepared as indicated.

	Volume	Final Concentration
ddH_2O	32.5–34 µL	-
10× Chelating buffer	4 µL	0.8×
$MgCl_2$	3–5 µL	1.5–2.5 mM
"Upstream" primer	0.5 µL	0.15 µM
Total volume	40 µL	

5. Amplify, store and analyze as described in Method A.

In Situ Polymerase Chain Reaction (ISPCR)

TISSUE PREPARATION

Five-micron-thick consecutive sections are cut and mounted on silane-coated (Oncor, Gaithersburg, MD, USA) or poly-L-lysine (PLL)-coated glass slides (Fisher Scientific, Pittsburgh, PA, USA). One tissue section must be placed in the center of each slide. Ideally, the size of the sections should be 0.5 to 1.0 cm^2. If using PLL-coated slides, baking the sections at 50°C for 2 to 4 h to increase the tissue adhesiveness is recommended. Cytological specimens can also be mounted on these slides and baking is not necessary. For cytological specimens, high cellular density is desirable.

The sections are deparaffined in xylene, rehydrated in decreasing graded ethanol and washed for 3 min in PBS. If they are not assayed the same day, they should be kept in paraffin at 4°C to avoid contamination or degradation. Optionally, enzyme pretreatment (i.e., proteinase K, DNase, RNase) can be performed on deparaffined tissue sections and stored at 4°C, if they will be assayed in a short period, i.e., the next day.

HIV ISPCR and HIV RT-ISPCR protocols take an average of three full working days and follow the same guidelines of other ISPCR methods. The same kits are used as for conventional PCR and RT-PCR. The technician schedule must be planned in advance. An ISPCR laboratory should have full-time and motivated staff.

HIV ISPCR Protocol

1. Deparaffined tissue sections or cytological samples are digested with 0.3 mg/mL of proteinase K solution, inside a humid chamber at 54°C for 15 to 20 min. **Note:** digestion time and temperature vary according to type of specimen and must be empirically verified.
2. Wash sections in ddH_2O for 10 min twice.
3. Place slides on a block at 95°C for 3 min to completely inactivate the enzyme.
4. Prehybridize the sections inside a humid chamber at 42°C for 30 min using 50 µL of a solution composed of 12% dextran sulfate, 2× SSC, 0.12 mM EDTA and 0.33 mg/mL salmon sperm DNA.
5. Remove the excess prehybridization solution from each section and heat each of them to 75°C in a slide thermocycler (MJ Research).
6. Slowly add 50 µL of PCR mixture to each section. This mixture consists of 33.75 µL ddH_2O, 5 µL 10× PCR Buffer II (final concentration [f.c.] 1×), 5 µL 25 mM $MgCl_2$ (f.c. 2.5 mM), 1 µL 10 mM each dNTP (f.c. 200 µM each), 1 µL 25 mM each primer, SK38/39 or SK102/432 (f.c. 0.5 mM each) and 0.25 µL 5 U/mL *Taq* DNA polymerase (f.c. 1.25 U/50 µL).
7. Cover each section with a glass coverslip one at a time and completely seal the edges with an appropriate amount of transparent nail polish. Avoid excessive nail polish, otherwise slides will not fit into the compartments of the thermocycler. Use the pipet tip to adjust the coverslips. Bubbles trapped beneath the coverslip will usually come out by themselves during heating. Therefore, do not try to remove them by pressing the coverslip. Keep all the slides at 75°C until the PCR mixture has been added to the last one.
8. Start the thermocycling. The temperature parameters for ISPCR are 95°C 1.5 min (initial denaturing), 30 cycles of 95°C (denaturing) 30 s, 55°C (annealing) 45 s and 72°C (extension) 30 s. This is followed by 72°C (final extension) 1.5 min and storage at 4°C (soaking).
9. The nail polish is softened with acetone and the coverslips are carefully removed with a surgical blade. Immediately, the sections are washed in 5× SSC, 2× SSC and PBS for 5 min each.
10. Bake the sections in an oven at 60°C for 20 min.
11. One hundred microliters of hybridization solution are added to each section. Heat to 95°C, 5 min for denaturing. Avoid evaporation of the solution. Hybridize overnight in a humid chamber at 45°C. The hybridization solution consists of 50% formamide, 25% dextran sulfate, 2× SSC, 0.33 mg/mL salmon sperm DNA, and HIV-1 biotinylated probe (SK19 or SK102 for SK38/39 and SK145/431 amplifications, respectively) at 250 to 400 pg/mL concentration.

12. Wash the sections with 5× SSC, 2× SSC and PBS for 5 min each.
13. Detection is performed using a kit for biotinylated probes (K600; DAKO, Carpenteria, CA, USA) based on the linkage of streptavidin and biotinylated alkaline-phosphotase coupled to NBT/BCIP colorimetric reaction (blue color). This reaction is carried out in darkness and monitored at about 15-min intervals under a light microscope, usually for no more than 1 h.
14. Once the detection is completed, wash the section in PBS for 5 min. If desired, slides may be slightly counterstained with Pyronin-Y, Nuclear Fast Red or Fast Green. These dyes should be dissolved in 2× SSC or PBS and not in ddH_2O.
15. Dry sections at 50°C in an oven and cover using permanent mounting media.

HIV RT-ISPCR

Method A. Reverse Transcriptase-Driven RT-ISPCR

1. Deparaffined tissue sections or cytological specimens are digested with 0.3 mg/mL proteinase K solution inside a humid chamber at 54°C for 15 to 20 min. Digestion process may be varied (see above).
2. Sections are washed twice in ddH_2O for 10 min and heated in a block at 95°C for 3 min to completely inactivate the enzyme.
3. Sections are pretreated with RNase-free DNase. Ten to twenty U/section incubating at 37°C for a minimum of 4 h. Overnight incubation is strongly advised to completely destroy the DNA.
4. The sections are extensively washed with several changes of ddH_2O for 20 min.
5. Ten microliters of RT mixture (GeneAmp RNA PCR Kit) are added to each section and incubated inside a humid chamber at room temperature for 15 min.

RT Mixture:

	Volume	Final Concentration
$MgCl_2$	2.0 µL	5 mM
10× Buffer II	1.0 µL	1×
ddH_2O	1.5 µL	-
dNTPs each	1.0 µL	1 mM
RNase inhibitor	0.5 µL	1 U/10 µL
Reverse transcriptase	0.5 µL	2.5 U/10 µL
Random hexamers	0.5 µL	2.5 µM
Total volume	10.0 µL	

6. Sections are incubated in a humid chamber at 42°C for 20 min.
7. Twenty microliters of prehybridization solution (same as used for HIV ISPCR) are added to each section.
8. Slides are placed in the slide thermocycler (MJ Research) set with one cycle at 99°C for 3 min (to inactivate the reverse transcriptase) and 5°C for 5 min.
9. Sections are incubated with the residual prehybridization solution in a humid chamber at 42°C for 20 min.
10. PCR mix is added to the sections, 40 µL/section.

PCR Mixture:

	Volume	Final Concentration
$MgCl_2$	2.0 µL	2 mM
10× PCR buffer	4.0 µL	1×
ddH_2O	37.75 µL	-
Taq DNA polymerase	0.25 µL	1.25 U/50 µL
Primer SK38	0.5 µL	0.25 µM
Primer SK39	0.5 µL	0.25 µM
Total volume	40 µL	

11. The amplification is performed in the slide thermocycler using the same parameters as for HIV ISPCR (see above).
12. Hybridization and detection are also performed as described in HIV ISPCR.

Method B. r*Tth* DNA Polymerase-Driven RT-ISPCR

1. Deparaffined tissue sections or cytological specimens are digested with 0.3 mg/mL proteinase K solution inside a humid chamber at 54°C for 15 to 20 min.
2. Sections are washed twice in ddH_2O for 10 min and heated on a block at 95°C for 3 min to inactivate the enzyme.
3. RNase-free DNase pretreatment is performed as described above.
4. Sections are extensively washed in ddH_2O for 20 min.
5. Twenty microliters of RT mixture (Thermostable r*Tth* Reverse Transcriptase RNA PCR Kit; Perkin-Elmer) are added to each section and incubated inside a humid chamber at 70°C for 25 min.

RT Mixture:

	Volume	Final Concentration
ddH_2O	11.5 µL	-
10× RT Buffer	2.0 µL	1×
$MnCl_2$	2.0 µL	1 mM
dNTPs each	0.4 µL	200 µM
r*Tth* DNA polymerase	2.0 µL	5 U/20 µL
“Downstream” primer SK39	1.0 µL	0.75 µM
Total volume	20.0 µL	

6. Next the humid chambers containing the slides are placed in a refrigerator at 4°C to stop the reaction.
7. Ten microliters of 12% dextran sulfate solution containing 1 mg of glycogen are added to each section.
8. Eighty microliters of PCR mixture are added to the sections.

PCR Mixture:

	Volume	Final Concentration
ddH_2O	61.0 μL	-
10× Chelating buffer	8.0 μL	0.8×
$MgCl_2$	10.0 μL	2.5 mM
"Upstream" primer SK38	1.0 μL	0.25 μM
Total volume	80.0 μL	

9. Amplification is performed in the slide thermocycler (MJ Research) using the same parameters as for HIV ISPCR (see above).
10. Hybridization and detection are also performed as described for HIV ISPCR.

REFERENCES

1. **Bagasra, O., S.P. Hauptman, H.W. Lischner, M. Sachs and R.J. Pomerantz.** 1992. Detection of human immunodeficiency virus type 1 provirus in mononuclear cells by *in situ* polymerase chain reaction. N. Engl. J. Med. *326*:1385-1391.
2. **Bagasra, O., T. Seshamma and R.J. Pomerantz.** 1994. *In situ* polymerase chain reaction: applications in the pathogenesis of diseases. Cell Vision *1*:48-51.
3. **Bagasra, O., T. Seshamma and R. Pomerantz.** 1993. Polymerase chain reaction *in situ*: Intracellular amplification and detection of HIV-1 proviral DNA and other specific genes. J. Immunol. Methods *158*:131-145.
4. **Bauer, H., Y. Ting, C. Greer et al.** 1991. Genital human papilloma virus infection in female university students as determined by PCR-based method. JAMA *265*:472-477.
5. **Brien, D., D. Billadeau and B. Van Ness.** 1994. RT-PCR assay for detection of transcripts from very few cells using whole cell lysates. BioTechniques *16*:586-590.
6. **Centers for Disease Control: Pneumocystis carinii pneumonia.** 1981. MMWR *30*: 250.
7. **Chiu, K.P., S.H. Cohen, D.W. Morris and G.W. Jordan.** 1992. Intracellular amplification of proviral DNA in tissue sections using the polymerase chain reaction. J. Histochem. Cytochem. *40*:333-341.
8. **Cullen, BR.** 1991. Regulation of HIV-1 gene expression. FASEB J. 5:2361-2368.
9. **Delgleish, A.G., P.Beverly, P.R. Clapham, D.H. Crawford, M.F. Greaves and R.A. Weiss.** 1984. The CD4 (T4) antigen is an essential component of the receptor for the AIDS retrovirus. Nature *312*:763-767.
10. **Donovan, R.M., H.C. Stuart, W.R. Peterson et al.** 1988. *In situ* detection of human immunodeficiency virus (HIV)-nucleic acid in H9 cells using nonradioactive DNA probes and an image cytophotometry system. J. Histochem. Cytochem. *36*:1573-1577.
11. **Gu, J.** 1994. Principle and application of *in situ* PCR. Cell Vision *1*:8-19.
12. **Haselt, W.A. and F. Wong-Staal.** 1988. The molecular biology of the AIDS virus. Sci. Am. *259*:52-62.
13. **Ho, D.D., R.J. Pomerantz and J.C. Kaplan.** 1989. Pathogenesis of infection with human immunodeficiency virus. N. Engl. J. Med. *321*:1621-1625.
14. **Imagawa, D.T., M.H. Lee, S.M. Wolinsky et al.** 1989. Human immunodeficiency virus type 1 infection in homosexual men who remain seronegative for prolonged periods. N. Engl. J. Med. *320*:1458-1462.
15. **Jimenez, E., E. Bache, M. Unger et al.** 1990. Immunohistochemical marker profiles of placentae of HIV-exposed pregnancies. Int. Conf. AIDS *6*:20-23.
16. **Kessler, H.A., J.A. Bick, J.C. Pottage, Jr., and C.A. Benson.** 1992. AIDS: Part 1. Dis. Mon. *38*:633-1990.
17. **Koenig, S., H.E. Gendelman, J.M. Orenstein et al.** 1986. Detection of AIDS virus in macrophages in brain tissue from AIDS patients with encephalopathy. Science. *233*:1089.
18. **Komminoth, P. and A. Long.** 1993. *In situ* polymerase chain reaction. An overview of methods, applications and limitations of a new molecular technique. Virchow Arch. [B] *64*:67-73.

19. **Komminoth, P., A. Long and H.J. Wolfe.** 1992. Comparison of *in situ* polymerase chain reaction (*in situ* PCR), *in situ* hybridization (ISH) and polymerase chain reaction (PCR) for the detection of viral infection in fixed tissue. Patologia. Suppl. *25*:253.
20. **Kwok, S., D.H.Mack and K.B.Mullis et al.** 1987. Identification of human immunodeficiency virus sequences by using *in vitro* enzymatic amplification and oligomer cleavage detection. J. Virol. *61*:1690.
21. **Laure, F., V. Courgnaud, C. Rozioux et al.** 1988. Detection of HIV-1 DNA in infants and children by means of the polymerase chain reaction. Lancet *2*:538-541.
22. **Maddon, P.J., A.G. Dalgleish, J.S. McDougal, P.R. Clapham, R.A. Weiss and R. Azel.** 1986. The T4 gene encodes the AIDS virus receptor and is expressed in the immune system and the brain. Cell *47*:333-348.
23. **Nuovo, G.** 1992. PCR *In Situ* Hybridization. Protocols and Applications. Raven Press, New York.
24. **Nuovo, G., A. Forde, P. MacConnell and R. Fahrenwald.** 1993. *In situ* detection of PCR amplified HIV-1 nucleic acids and tumor necrossis factor cDNA in cervical tissues. Am. J. Pathol. *143*:40-48.
25. **Nuovo, G., F. Gallery and P. MacConnell.** 1993. Importance of different variables for optimizing *in situ* detection of PCR-amplified DNA. Amplifications *11*:4-6.
26. **Nuovo, G., G. Gorgone, P. MacConnell, M. Margiotta and P. Gorevic.** 1993. *In situ* localization of PCR-amplified human and viral cDNAs. PCR Meth. Appl. *2*:117-123.
27. **Nuovo, G., M. Margiotta, P. MacConnell and J. Becker.** 1992. Rapid *in situ* detection of PCR-amplified HIV-1 DNA. Diagn. Mol. Pathol. *1*:98-102.
28. **Ou, C.-Y., S. Kwok, S.W. Mitchell et al.** 1988. DNA amplification for direct detection of HIV-1 in DNA of peripheral blood mononuclear cells. Science *239*:295-297.
29. **Patterson, B., M. Till, P. Otto et al.** 1993. Detection of HIV-1 DNA and messenger RNA in individual cells by PCR-driven *in situ* hybridization and flow cytometry. Science *260*:976-979.
30. **Peterson, A. and B. Seed.** 1988. Genetic analysis of monoclonal antibody and HIV binding sites on the human lymphocyte antigen CD4. Cell *54*:65-72.
31. **Qinn, T.C.** 1990. The epidemiology of the human immunodeficiency virus. Ann. Emerg. Med. *19*:225-232.
32. **Rappaport, J., S.J. Lee, K. Khailili and F. Wong-Staal.** 1989. The acidic amino-terminal region of the HIV-tat protein constitutes an essential activating domain. New Biol. *1*:101-110.
33. **Rogers, M.F., C.-Y. Ou, M. Rayfield et al.** 1989. Use of the polymerase chain reaction for early detection of the proviral sequences of human immunodeficiency virus in infants born to seropositive mothers. N. Engl. J. Med. *320*:1649-1654.
34. **Sardelli, A.** 1993. Plateau effect – Understanding PCR limitations. Amplifications (Perkin-Elmer) *9*:1-3.
35. **Shaw, G.M., M.E. Harper, B.H. Han, L.G. Epstein et al.** 1985. HTLV-III infection in brains of children and adults with AIDS encephalopathy. Science *227*:177.
36. **Shibata, D.K.** 1992. The polymerase chain reaction and the molecular genetic analysis of tissue biopsies. p. 85-111. *In* C.S. Herrington and J.O. McGee (Eds.), Diagnostic Molecular Pathology: A Practical Approach, Vol. 2. Oxford University Press, Oxford.
37. **Smith-Norowitz, T., E. Zevallos, V. Anderson, H. Durkin and E. Bard.** 1994. A modified *in situ* PCR-hybridization procedure. Cell Vision *1*:84.
38. **Stanta, G. and C. Schneider**. 1991. RNA extracted from paraffin-embedded human tissues is amenable to analysis by PCR amplification. BioTechniques *11*:304-308.
39. **Webner, J.N. and R.A. Weiss.** 1988. HIV infection: The cellular picture. Sci. Am. *259*:101-109.
40. **Wiley, C.A., R.D. Schrier, J.A. Nelson, P.W. Lampert and M.B. Oldstone.** 1986. Cellular localization of human immunodeficiency virus infection within the brains of acquired immune deficiency patients. Proc. Natl. Acad. Sci. *8*:7089
41. **Zehbe, I., G. Hacker, E. Rylander, J. Sällström and E. Wilander.** 1992. Detection of single HPPV copies in SiHa cells by *in situ* polymerase chain reaction (*in situ* PCR) combined with immunoperoxidase and immunogold-silver staining (IGSS) techniques. Anticancer Res. *12*:2165-2168.
42. **Zevallos, E., V. Anderson, E. Bard and J. Gu.** 1994. Detection of HIV-1 sequences in placentas of HIV-infected mothers by *in situ* PCR. Cell Vision *1*:116-122.
43. **Zevallos, E., E. Bard, V. Anderson, N. Carson and J. Gu.** 1994. HIV *in situ* PCR and reverse transcribed *in situ* PCR. Cell Vision *1*:52-54.

44. **Zevallos, E., E. Bard, V. Anderson and J. Gu.** 1994. An *in situ* PCR study of HIV-1 infection of lymphoid tissues and peripheral lymphocytes. Cell Vision *1*:87.

45. **Zevallos, E., V. Anderson, E. Bard, T. Smith-Norowitz, M. Nowakowski and J. Gu.** 1994. Reverse transcribed *in situ* RT-ISPCR of retinoblastoma mRNA. Cell Vision *1*:88.

46. **Zevallos, E., V. Anderson, M. Nowakowski, E. Bard and J. Gu.** 1994. Detection of human retinoblastoma gene expression by rTth-driven reverse transcribed *in situ* PCR. Cell Vision *1*:88.

47. **Zevallos, E., V. Anderson, M. Nowakowski, E. Bard and J. Gu.** 1994 Detection of HIV-1 RNA in lung and bronchioloalveolar lavage by reverse transcribed *in situ* PCR. Cell Vision *1*:87.

48. **Zevallos, E., V. Anderson, M. Nowakowski, E. Bard and J. Gu.** 1994 Detection of HIV-1 sequence in lung and bronchioloalveolar lavage by *in situ* PCR. Cell Vision *1*:87.

Address correspondence to Jiang Gu, Deborah Research Institute, Trenton Road, Browns Mills, NJ 08015, USA.

ILLUSTRATIVE EXAMPLES *(Color reproductions of these figures appear on p. 140-143.)*

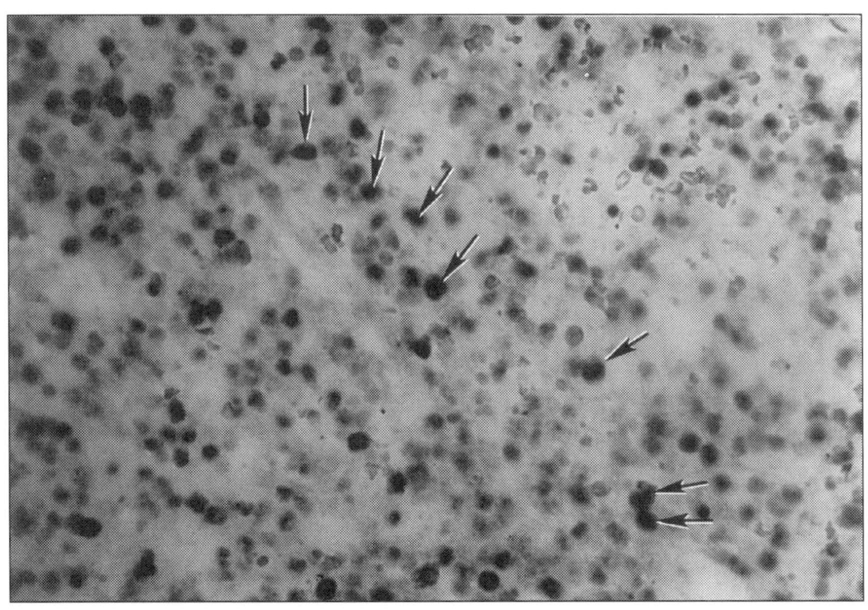

Figure 1. HIV-1 ISPCR. Lymph node section from a pediatric AIDS autopsy with lymphoproliferative syndrome. The *in situ* amplification was accomplished using HIV-1 *gag* primers SK38 and SK39, as described in the HIV ISPCR protocol above. *In situ* hybridization was performed using biotinylated HIV-1 SK19 probe. The majority of lymphocytes show nuclear positivity (blue color). Some positive cells have enlarged nuclei (arrowheads). Nuclear Fast Red counterstain. Magnification, ×20.

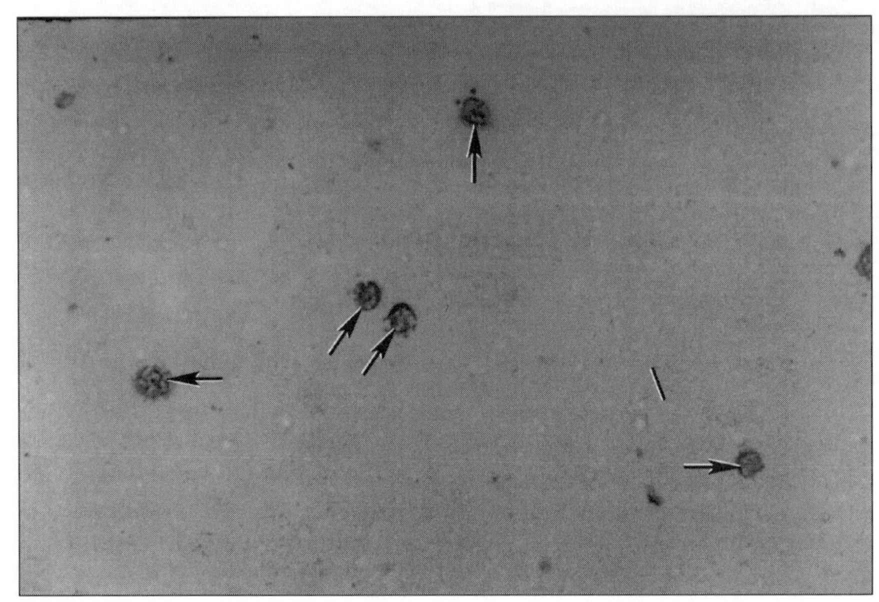

Figure 2. HIV-1 RT-ISPCR. Peripheral blood mononuclear cells from AIDS patient were obtained by Ficoll®-gradient centrifuge and cytospin. The reverse transcription and amplification were achieved using the r*Tth* DNA polymerase Driven RT-ISPCR method with HIV-1 *gag* primers SK38 and SK39. *In situ* hybridization was performed with biotinylated HIV-1 SK19 probe. Lymphocytes (arrowheads) display cytoplasmic positivity (blue color). Nuclear Fast Red counterstain. Magnification, ×40.

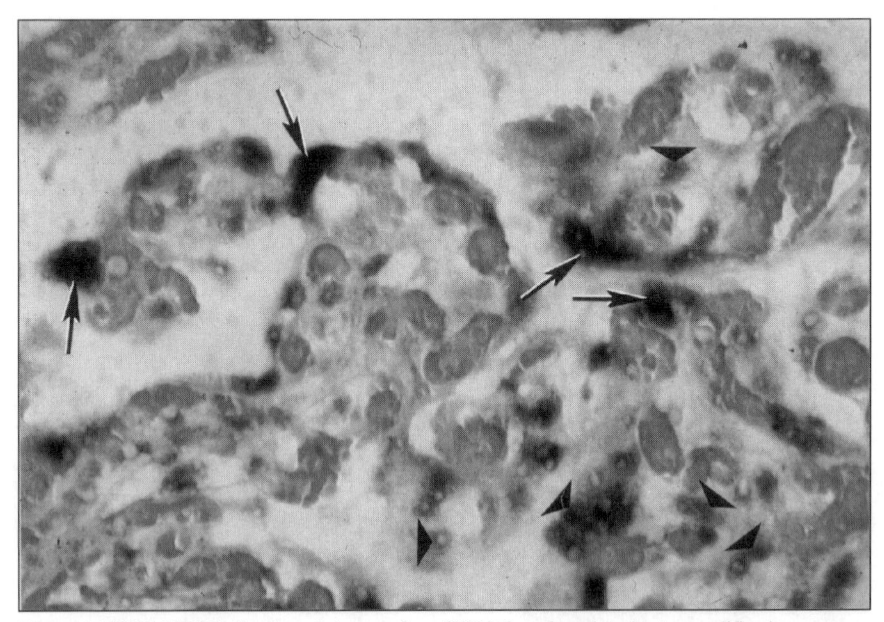

Figure 3. HIV-1 ISPCR. Section of placenta from HIV-infected mother. *In situ* amplification was performed with HIV-1 *gag* primers SK145 and SK431, as described in the HIV ISPCR protocol. *In situ* hybridization was carried out using biotinylated HIV SK102 probe. Syncitotrophoblast (arrows) and Hofbauer cell (arrowheads) display nuclear positivity (blue color). These placental cells are frequently the most positive. Nuclear Fast Red counterstain. Magnification, ×40.

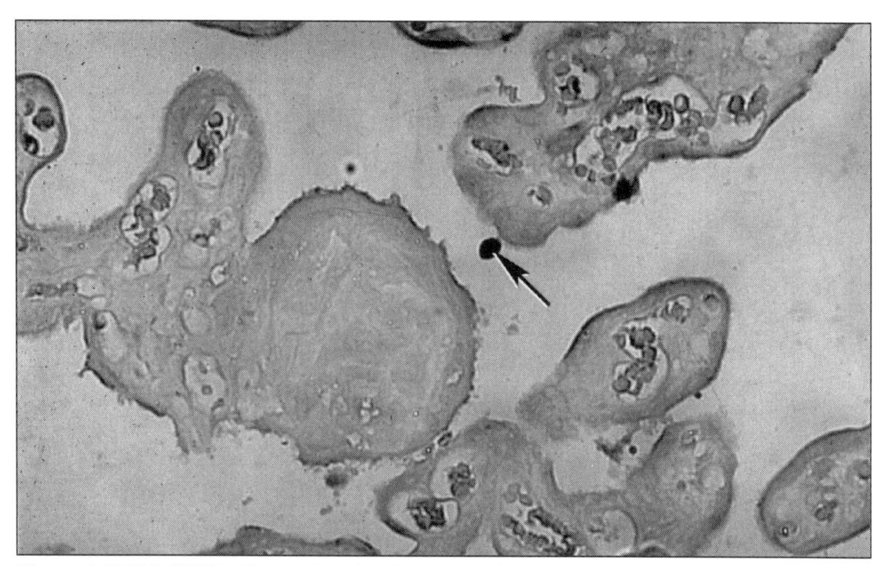

Figure 4. HIV-1 ISPCR. Placental section from HIV-infected mother. *In situ* amplification was performed using HIV-1 *gag* primers SK38 and SK39, and *in situ* hybridization and detection were carried out according to the described HIV ISPCR protocol. The arrowhead points to a positive lymphocyte (dark blue) of the intervillous space in close contact with the trophoblastic layer. Cell-to-cell interaction as demonstrated here may be an important route in the pathogenesis of vertical transmission of HIV-1. Fast Green counterstain. Magnification, ×20.

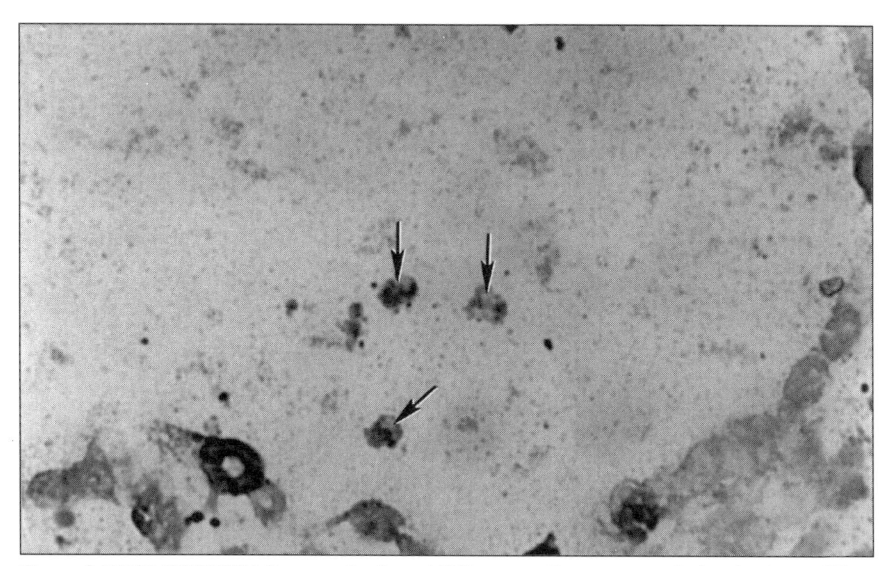

Figure 5. HIV-1 RT-ISPCR. Lung section from AIDS autopsy. Reverse transcription, *in situ* amplification, hybridization and detection were performed according to the Reverse Transcriptase Driven RT-ISPCR protocol. The initial reverse transcription was achieved with random hexamers. The *in situ* amplification was performed by using HIV-1 *gag* primers SK38 and SK39. Arrowheads point to cytoplasmic positivity (blue color) almost exclusively in intra-alveolar cells with macrophage morphology. Although there is significant distortion, some faint positivity is seen in other intra-alveolar cells. The relevance of this finding is that intra-alveolar macrophages may be the sites of viral replication indicating an "active" infection. Nuclear Fast Red counterstain. Magnification, ×40.

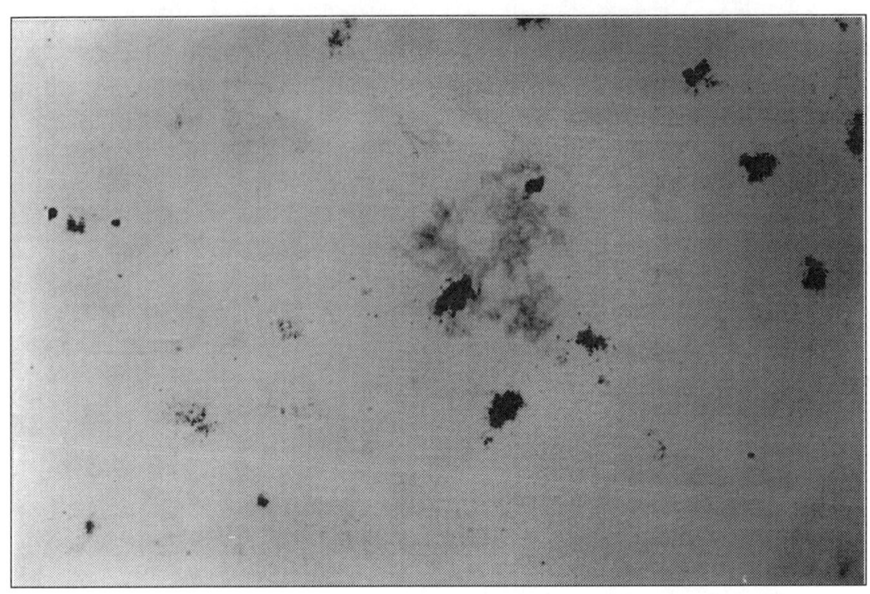

Figure 6. HIV-1 RT-ISPCR. Bronchioloalveolar lavage (BAL) from AIDS patient. BAL cells were cytospun on glass slides. Reverse transcription, *in situ* amplification, hybridization and detection were performed as previously described. Moderate heat distortion is observed. Strong cytoplasmic positivity (dark blue) is present in cells with macrophage morphology. The pink amorphous structure in the center corresponds to epithelial cells. Pyronin Y counterstain. Magnification, ×40.

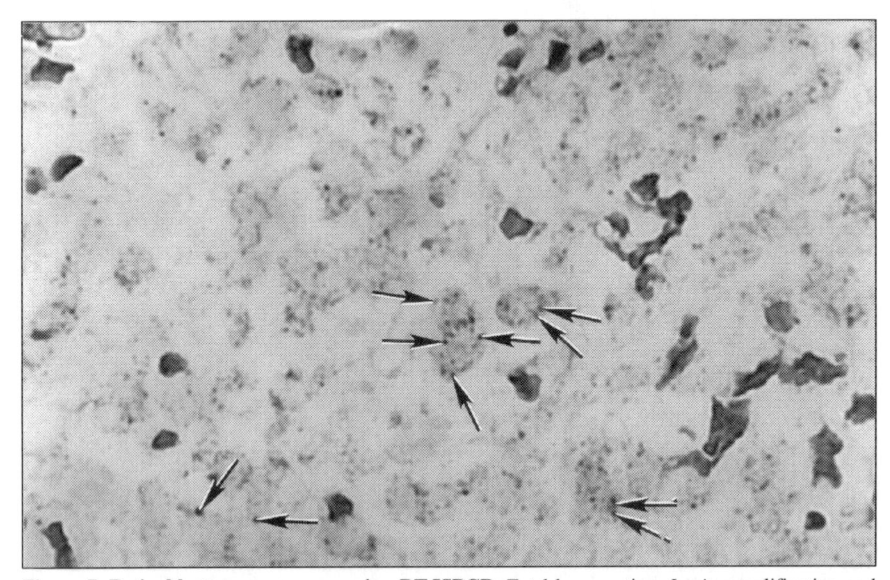

Figure 7. Retinoblastoma gene expression RT-ISPCR. Fetal lung section. *In situ* amplification and detection were performed according to the r*Tth* DNA Polymerase Driven RT-ISPCR protocol. Sections were pretreated with RNase-free DNase overnight. Specific retinoblastoma complementary primers for mRNA. Biotin-dUTP were added to the PCR mixture for direct labeling of amplified products. Granular positivity is restricted to cytoplasmic location (arrowheads) in most of the cells. No nuclear positivity is observed, which indicates absence of "pseudo-primer false positivity" (see text). Direct labeled ISPCR with specific RNA complementary for commonly expressed genes can be useful to verify amplifiability of mRNAs. Magnification, ×40.

Oligonucleotide-Primed *In Situ* Transcription and Immunogold-Silver Staining Systems: Localization of mRNA in Tissues and Cells

Lawrence E. De Bault[1] and Bao-Le Wang[2]

[1]Department of Pathology, University of Oklahoma Health Sciences Center, Oklahoma City, OK and [2]Bligh Cancer Research Laboratories, Finch University of Health Sciences/Chicago Medical School, North Chicago, IL, USA

SUMMARY

The essence of in situ *transcription (IST) is the synthesis of complementary DNA (cDNA) within a cell. In its simplest form, IST is performed by annealing a specific primer to its complementary mRNA in a tissue section or individual cell, washing away the unhybridized primer and synthesizing cDNA in the presence of reverse transcriptase and deoxynucleotide triphosphates. Protocols are presented that take advantage of the simplicity of IST, the use of digoxigenin-labeled dUTP and an immunochemical bridge, and the resolution and high contrast of immunogold-silver staining (IGSS). The fixation and handling of the tissues and cells, and the IST and IGSS procedures were adapted to allow the localization of γ-glutamyl transpeptidase (γ-GTP) and its mRNA in rat tissue, i.e., γ-GTP protein by immunocytochemistry, γ-GTP enzyme activity by enzymecytochemistry and γ-GTP mRNA by IST and IGSS.*

INTRODUCTION

The purpose of this chapter is to briefly review the historic background leading to the development of *in situ* transcription (IST) and a closely related technique, primer *in situ* labeling (PRINS), state the principles of IST in the context of *in situ* hybridization (ISH) and *in situ* polymerase chain reaction (IS-PCR) (in their various forms), point out the advantages of IST, particularly when coupled with an immunocytochemical detection system such as immunogold-silver staining (IGSS), and give examples of IST application to aldehyde-fixed paraffin-embedded tissue and to cells grown in tissue culture with specific reference to the detection and localization of γ-glutamyl transpeptidase (γ-GTP) mRNA. This chapter is not an exhaustive treatment of IST applications to fixed tissues and cells because the technique is relatively new, the literature is sparse and the experience with the technique in the scientific community is limited. Other aspects and details of ISH, conventional PCR, IS-PCR and their various forms are addressed elsewhere in this volume and will not be covered in this chapter.

Historic Background

Advancement in scientific discovery and knowledge and the emergence of new fields or subfields of investigation are preceded by key discoveries and innovations. From our perspective, there were about a dozen or so historic milestones spanning approximately two decades that made possible the application of IST to tissues and cells *in situ*. The field was started by four somewhat unrelated discoveries: the 1969 work of Gall and Pardue (26), demonstrating the formation and detection of RNA-DNA hybrid molecules *in situ;* the 1970 work of Casperson et al. (10), demonstrating the Q-band staining of chromosomes; the 1970 discovery of RNA-dependent DNA polymerase (or reverse transcriptase) by two independent groups, Baltimore (2) and Temin and Mizutani (67); and the 1971 work of Kleppe et al. (43), demonstrating the replication of short synthetic DNA segments catalyzed by DNA polymerases. The field picked up steam with three landmark discoveries: the 1973 work of Cohen et al. (13), demonstrating the cloning of DNA fragments into plasmid vectors; the 1975 work of Kohler and Milstein (49), demonstrating the development and *in vitro* production of monoclonal antibodies; and the 1976 work of Chien et al. (12), discovering thermostable DNA polymerase. The next major advance did not occur until the next decade, in 1981, when Langer et al. (50) enzymatically synthesized biotin-labeled polynucleotides. This was followed by the work of Saiki et al. in 1985 (63) that achieved *in vitro* primer-mediated amplification of genomic DNA. In 1987 Kogan et al. (48) introduced the use of thermostable DNA polymerase and made possible the practical use of the polymerase chain reaction (PCR).

The aforementioned milestones were essential advances for the advent of IST or were indirectly necessary for the proof of the principles of the IST technique. IST was first described and applied to tissue sections by Tecott et al. in 1988 (66) and confirmed and applied to cells *in vitro* by Longley et al. in 1989 (57).

Principle of IST Method

IST is based on the hybridization of a primer (usually a specific oligonucleotide) to mRNA in a tissue section or cell preparation on a glass slide as is done in conventional *in situ* hybridization (ISH) (14,34,42,53–56). Transcription of the mRNA is achieved by adding reverse transcriptase and labeled nucleotides that allow the synthesis of a labeled complementary DNA (cDNA). Since the synthesized cDNA remains associated with its mRNA template as a cDNA-mRNA hybrid, anatomical distribution and cellular localization are preserved.

Development of IST Method

The technique of IST was first described in 1988 by Tecott et al. from Stanford University School of Medicine (66) and was applied to tissue sections. These investigators used an oligonucleotide complementary to proopinmelanocortin (POMC) mRNA as a primer for the specific synthesis of cDNA. This was done on fresh-frozen paraformaldehyde-fixed sections of rat pituitary using reverse transcriptase and radiolabeled ^{3}H-dCTP in the cDNA elongation process, allowing for the incorporation of the label at an anatomically defined site. The newly synthesized cDNA was visualized in the tissue sections by autoradiography. In 1989, Longley et al. from Yale University School of Medicine (57) described the application of IST to *in vitro* disaggregated and isolated epidermal cells centrifuged onto glass slides and fixed in paraformaldehyde. These investigators used an oligonucleotide primer specific to the alpha-2 domain of CD1a mRNA, and ^{35}S-dCTP incorporation into cDNA followed by autoradiography to visualize the CD1a mRNA positive cells.

In the original works (57,66), radio labels were used, and the localization of the labeled cDNA was detected by autoradiography. Nonradioactive detection systems have been used in ISH and can be applied to IST, i.e., fluorochrome-labeled deoxynucleotide triphosphates (dNTPs) in fluorescence methods (3,4,44,55,56,68,71), biotin-labeled (28,35,42,45,47,50,55,59,60) or digoxigenin (DIG)-labeled (18,19, 33,35,61) dNTPs in immunochemical methods.

A technique closely related to IST for mRNA is primed *in situ* labeling (PRINS), which was first used by a Danish group to localize chromosomal DNA (44–47) and was later confirmed by a group in Scotland (28,59). For clarity, the PRINS technique applied to DNA will be referred to as PRINS-DNA. The PRINS-DNA procedure uses unlabeled DNA probes or oligonucleotides as the primer, and DNA polymerase (Klenow or *Taq*-I) and biotin or DIG-labeled dNTPs to synthesize labeled DNA *in situ*. The site of synthesis was detected by immunocytochemistry using fluorochromes as reporter molecules. A variation of PRINS-DNA can be used in the detection of mRNA (46,60) and, when applied to mRNA, will be referred to as PRINS-mRNA. PRINS-mRNA is virtually the same as IST in that it uses an unlabeled primer complementary to a specific mRNA sequence, and reverse transcriptase and labeled dNTPs to synthesize a labeled cDNA. Thus all comments made about IST should also apply to PRINS-mRNA.

Advantages of IST

IST has several attractive advantages. Since chain elongation is independent of the length of the primer, oligonucleotide primers induce as much (if not more) labeling of the target mRNA than a much longer pre-labeled probe, thus increasing the sensitivity of the method. Another advantage of IST is its application to the detection of minor sequence variations in RNA by the proper selection of the sequence; thus the position of the primer can control whether or not there will be chain elongation. A significant technical advantage of the method is that the probe (primer) is unlabeled and labeling occurs only secondarily to specific hybridization; also, the unincorporated labeled nucleotide can be washed away easier, which results in a lower background. In addition, fewer procedural steps allows for a shorter cumulative incubation time, resulting in less degradation in the tissue morphology that often accompanies ISH or ISPCR. Another advantage of the milder conditions of IST is that it allows for the detection of mRNA in cell suspension intended for flow cytometry without clumping and disintegration of cells (1). The application of IST to the detection of minor sequence variations should be superior to ISH in that it is well known that the last few nucleotides of the primer are crucial to the initiation of chain elongation (7).

Applications

Some of the early applications of IST to mRNA were on fixed tissue sections or cells where the reporter molecules remained *in situ*, while other applications used IST as a first step, followed by other *in vitro* procedures for the detection of the synthesized cDNA. For example, IST was used to investigate the temporal expression of mRNAs in developing embryos (24,25) where the embryos were fixed in Bodian's fixative (80% ethanol, 37% formaldehyde and 5% glacial acetic acid) or 4% paraformaldehyde and processed by standard histological procedures.

IST has been used in the analysis of gene expression in live cells where mRNA in single live neurons *in vitro* was injected with primer, dNTPs and reverse transcriptase via whole cell patch electrode, the contents of the cell harvested by suction back into the electrode, the electrode incubated *in vitro* where the cDNA is first synthesized by IST followed by replication of cDNA to many copies of amplified RNA (aRNA) (69). The aRNA was assessed by Southern and Northern blot analysis. IST was also used to produce cDNA from low abundance mRNA, which was detected directly or was further amplified before analysis (20,21,30,38).

IST has been used to assess the transcriptional regulation of pharmacological agents at the mRNA level by investigation of mRNA structure *in situ* in fixed cells and tissues by synthesizing the cDNA primed by an oligonucleotide complementary to a putative POMC stem-loop structure (22). The IST cDNA transcript was extracted from the section by alkaline denaturation of the mRNA-cDNA hybrids and electrophoresed on denaturing DNA sequencing gels followed by autoradiography. The resulting banding patterns indicated that pharmacological agents which modulate POMC translation also produced local changes in the banding patterns.

IST has been used to localize γ-GTP mRNA in paraffin sections, from rat kidney (70) and from cell cultures of rat brain microvessels (19).

Example of IST Application to Tissues and Cells: Recommended Protocols for the Localization of γ-GTP

In this set of protocols, we present an adaptation of the IST technique by combining *in situ* DIG labeling and IGSS systems to demonstrate and localize γ-GTP mRNA in rat kidneys. These protocols have been published in part elsewhere (18,19,70). Enzymecytochemistry and immunocytochemistry have been used to demonstrate that the predominant localization of γ-GTP in rat kidney is on the luminal surface brush border of proximal tubule epithelium. The principal objectives of these protocols are: 1) to accommodate the procedures to paraffin-embedded tissue sections; 2) to substitute an alternative method for the alkaline phosphatase detection system that often gives false positive results in kidney, especially on epithelial cell brush borders; and 3) to study the spatial relationship between γ-GTP's protein and its activity, and γ-GTP mRNA.

ABSTRACT OF PROTOCOL(S)

Animals were perfuse-fixed with 3% paraformaldehyde and 0.5% glutaraldehyde in 0.1 M phosphate-buffered saline (PBS) (pH 7.4), and the kidney tissue embedded in paraffin by standard procedures. Under RNase-free conditions, three 25-nucleotide oligonucleotide probes complementary to γ-GTP mRNA were hybridized separately or as a mixture to 5-μm sections, overnight. The primed γ-GTP mRNA was then transcribed *in situ* by incubation with a DIG-labeled dUTP mixture containing 0.1 mM dATP, 0.1 mM dCTP, 0.1 mM dGTP, 0.065 mM dTTP, 0.035 mM DIG-dUTP, and avian myeloblastosis virus (AMV) reverse transcriptase at pH 7.4, 42°C, for 60 min. After washing the *in situ* transcribed sections, the incorporated DIG was bridged with sheep anti-DIG IgG and detected with 10 nm gold-conjugated rabbit anti-sheep IgG followed by silver enhancement. Controls consisted of the omission of the oligonucleotide probes or the substitution of unrelated 25-nucleotide oligonucleotides in the hybridization step.

A. Preparation of Rat Kidney for IST and Cytochemistry of γ-GTP and Its mRNA

Animals: Two-month-old Wistar male rats weighing between 150 and 200 g were used.

Fixative: 3% paraformaldehyde + 0.5% glutaraldehyde + 2.5% sucrose in 0.1 M PBS, pH 7.4.

Handling Animals: Animals are housed in accredited animal facilities and fed standard rat chow for 1 week prior to experimentation. Animals used for tissue source were anesthetized with ethyl ether to light ether narcosis.

Perfusion Fixation: All reagents used and steps in the perfusion and fixation procedure were performed at +4°C, unless otherwise stated (39,51).

1. 0.1 M PBS pH 7.4 was perfused through the left ventricle for 5 min.
2. Fixative was perfused by the same route for 10 min.
3. Kidneys were removed and samples cut for subsequent processing (cryosectioning, paraffin embedding and/or plastic embedding for electron microscopy).

Immersion Fixation and Processing: The cut samples were immersed in the same fixative according to the protocols below:

a. For Cryosections: The samples are immerse-fixed for an additional 80 min followed by:

   i. Wash 2 times with 0.1 M PBS pH 7.4, 10 min each.
   ii. Infiltrate with 0.5 M sucrose in 0.1 M PBS pH 7.4, overnight.
   iii. Embed in O.C.T. compound (Tissue Teck II, Miles, Elkhart, IN, USA) and freeze in liquid nitrogen.
   iv. Store at -45°C or lower until cryosectioned (40).

b. For Paraffin Embedding: The samples are immerse-fixed for an additional 150 min followed by:

   i. Wash 5 times with 0.1 M PBS pH 7.4, 15 min each.
   ii. Dehydrate in an ethanol series, 30% for 20 min, 70% for 30 min, 95% for 90 min at room temperature.

iii. Infiltration was performed by machine processing and included 2 changes of 95% ethanol for 20 min each, 3 changes of 100% ethanol for 15, 20 and 30 min, respectively, 2 changes of xylene for 20 and 30 min respectively, all at 40°C, and 2 changes of paraffin for 45 min each at 57°C. The tissue is embedded in flat molds (58,64).

Note: Completion of the entire embedding process on the same day that the perfusion fixation is performed is important in maintaining maximum antigenicity and mRNA reactivity.

c. For EM: The samples are immerse-fixed for an additional 90 min followed by:

i. Wash with 0.1 M PBS pH 7.4, overnight.
ii. Dehydrate in an ethanol series at room temperature, 30% for 20 min, 70% for 30 min, 95% for 90 min.
iii. Infiltrate with 2 changes of L.R. White (Ernest F. Fullam, New York, USA) 1 h each at room temperature.
iv. Embed in L.R. White and cure at 56°C for 40 h according to L.R. White Data Sheet (15–17,27,29).

B. IST of γ-GTP mRNA in Rat Kidney Paraffin Sections

Paraffin Embedding: Kidneys were paraffin-embedded according to method outlined above.

Sectioning: 5-μm sections were floated on a 0.1% DEPC- (Diethyl Pyrocarbonate; Sigma Chemical, St. Louis, MO, USA) treated water bath at 42°C and picked up on silanized slides (62). **Note:** It is important to perform the *in situ* transcription and immunogold-silver staining procedures immediately after cutting; cut sections stored for days or weeks lose reactivity and background often increases.

Deparaffinization and Rehydration: Before deparaffinization, the slides were incubated in a 60°C oven for 20 min. The following steps were conducted at room temperature:

1. Deparaffinize with 3 changes of xylene for 5 min each, followed by 3 changes of 100% ethanol for 3 min each.
2. Rehydrate in ethanol series, 95%, 75%, 50% for 3 min each followed by 3 changes of 2× standard saline solution (SSC) in 0.1% DEPC-treated distilled water for 10 min each. (2× SSC solution consists of 0.15 M NaCl, 0.015 M sodium citrate, and adjusted to pH 7.0 with 1 *N* HCl)

In Situ Hybridization (ISH) of Oligonucleotide Primer: All reagents were prepared on ice and used at the indicated temperatures (41).

3. Each slide is incubated with 20 μL of ISH mixture containing oligonucleotide primer(s) (covered with a glass coverslip) at 42°C overnight, followed by 30-min incubation at room temperature. The final mixture contained:

   50% formamide

   4× SSC

   0.02% bovine serum albumin (BSA)

5 mM DTT (dithiothreitol)

0.6 U RNase inhibitor/μL

10 μM oligonucleotide (see recipe for ISH solution below)

Control slides were incubated with the same mixture but with an unrelated oligonucleotide substituted, or this step (#3) was omitted.

4. Wash 2 times with 2× SSC and 2 times with 0.5× SSC, 15 min each at room temperature.
5. Hold slides in 0.5× SSC for 2 h before proceeding with IST.

In Situ Transcription (IST) of γ-GTP mRNA: All reagents were prepared on ice and used at the indicated temperatures (60,66).

6. Each slide was incubated with 15 μL of IST mixture containing DIG-labeled dUTP (covered with a glass coverslip) at 37°C for 60 min followed by 45°C for 10 min. The final mixture contained:

   50 mM Tris-HCl, pH 7.4

   6 mM $MgCl_2$

   40 mM KCl

   5 mM DTT

   0.02% BSA

   0.1 mM DIG-DNA labeling mixture

   0.6 U RNase inhibitor/μL

   1 U AMV reverse transcriptase/μL (see recipe for IST Solution below)

   In negative control slides the procedure was modified by: a) omitting the ISH oligonucleotide step #3, b) omitting the DIG-DNA labeling mixture from IST step #6, c) omitting the AMV reverse transcriptase from IST step #6, or d) a combination of these omissions.

7. Wash 2 times with 2× SSC 30 min each at room temperature.
8. Wash 2 times with 0.05× SSC 1 h each at 35°C.
9. Wash 3 times with 0.05 M Tris-buffered saline (TBS) pH 7.4 containing 0.25% Triton® X-100 10 min each at room temperature.

Immunocytochemical Detection of Digoxigenin: All immunoreagents were diluted with 1% BSA in 0.05 M TBS pH 7.4 (52,66,73).

10. Slides were preincubated in 1% BSA in 0.05 M TBS for 15 min at room temperature.
11. Slides were incubated with 50 μL of 1° antibody mixture at +4°C overnight, followed by 1 h incubation at room temperature. The final mixture contained sheep anti-DIG antibody diluted 1:50 in 1% BSA in 0.05 M TBS.
12. Wash 3 times with 0.05 M TBS 10 min each followed by wash with 0.02 M TBS pH 8.2 containing 0.25% Triton X-100 for 10 min at room temperature.
13. Incubate in 1% BSA in 0.02 M TBS pH 8.2 for 15 min at room temperature.
14. Incubate with 50 μL of 2° antibody mixture for 1 h at room temperature. The final mixture contained 10 nM gold-conjugated rabbit anti-sheep IgG diluted 1:10 in 1% BSA in 0.02 M TBS.

15. Wash once with 1% BSA in 0.02 M TBS, pH 8.2 for 10 min, followed by 3 washes of 0.05 M TBS, 10 min each.

Silver Enhancement of 10 nM Gold:

16. Wash with distilled water.
17. Incubate in silver enhancement mixture for about 20 min at 22°C. The final mixture contained (32,36):

    5.5 mM silver lactate

    77 mM hydroquinone

    120 mM citric acid

    80 mM sodium citrate

    10% gum arabic (see recipe for silver enhancement solution below)
18. Wash with distilled water 5 times 2 min each.
19. Counterstain lightly with hematoxylin and eosin (H&E) (optional).
20. Mount in Permount® (Fisher Scientific, Pittsburgh, PA, USA) and coverslip.

C. Support Protocols

Preparation of *In Situ* Hybridization (ISH) Solution: Prepare solution with cold reagents and hold on ice until used. Twenty microliters of final working ISH solution is needed for each slide.

	Formulation		Final Concentration
Solution #1	Deionized formamide	500 µL	50% formamide
	20× SSC	200 µL	4× SSC
	10 mg/10 mL BSA	200 µL	200 µg BSA/mL
	38.5 mg/5 mL DTT	100 µL	5 mM DTT
	Total	1 mL	

Solution #2 Take 160 µL of solution #1 and add 2.5 µL of 40 U/µL RNase inhibitor (Boehringer Mannheim, Indianapolis, IN, USA) to give a total of 100 units in 162.5 µL. (Enough for 8 slides.) (5–8)

Solution #3 Final working solutions: Makes 44 µL to 46 µL of ISH mixture containing oligonucleotide — enough for 2 slides when used at 20 µL/slide. Adjust volume for additional slides.

Oligonucleotide A = 40 µL solution #2 + 4 µL Oligo-A
Oligonucleotide B = 40 µL solution #2 + 4 µL Oligo-B
Oligonucleotide C = 40 µL solution #2 + 4 µL Oligo-C
Oligonucleotide mixture = 40 µL solution #2 + 2 µL Oligo-B
+ 2 µL Oligo-B + 2 µL Oligo-C

<u>Preparation and Synthesis of Oligonucleotides and Solutions</u>: Three oligonucleotides were synthesized by the Molecular Biology Resource Facility at the University of Oklahoma Health Sciences Center, Oklahoma City, OK, USA. They were synthesized on an Applied Biosystems Model 380B or 394A (Foster City, CA, USA) according to the β-cyanoethyl phosphoramidite chemistry method (9).

The oligonucleotides were purified by reverse-phase high-pressure liquid chromatography on a 4.6- × 250-mm C18 column (Rainin Instrument, Emeryville, CA, USA). The column was equilibrated with 0.02 M triethylammonium acetate, pH 7.0; and the elution was accomplished by a linear gradient of 5% to 30% acetonitrile in 12 min.

The three 25-nucleotide oligonucleotides complementary to segments of γ-GTP mRNA were:

1) 5′> CCCCCGATGCCCATACTGTGGGCAT <3′	100 pM/μL
2) 5′> AGGGTTGGAAGAGGCGAGCCCAGGG <3′	100 pM/μL
3) 5′> TGGAGCCAAAGTAGAGGTTGATGGT <3′	100 pM/μL

<u>Preparation of *In Situ* Transcription (IST) Solution</u>: Prepare solution with cold reagents and hold on ice until used. Fifteen microliters of final working IST solution is needed for each slide.

Formulation		Final Concentration
Solution #1		
0.05 M Tris-HCl pH 6.5	40 mL	
$MgCl_2 \bullet 6H_2O$ (MW:203.3)	48 mg	6 mM
KCl (MW:74.55)	120 mg	40 mM
DTT (MW:154.3)	31 mg	5 mM
BSA	8 mg	200 μg/mL
Total	40 mL	

Solution #2		
<u>Final Working Solution</u>:		
Solution #1	140.0 μL	
BM* Digoxigenin-DNA labeling mixture (10×)	15.0 μL	approx. 0.1 mM/base
BM RNase Inhibitor (40 U/μL)	2.5 μL	100 U/163.5 μL
BM AMV reverse transcriptase (24 U/μL)	6.0 μL	150 U/163.5 μL
Total	163.5 μL	

* = Boehringer Mannheim (11,23,31,37,65,72)

Preparation of Silver Enhancement Solution: Makes 100 mL of final working solution.

Solution #1

Citric acid, monohydrate	2.55 g	120 mM (concentration in final solution)
Sodium citrate, dihydrate	2.35 g	80 mM
Distilled H_2O	50 mL	
Add 50% gum arabic (in H_2O)	20 mL	10%
Subtotal	70 mL	

Solution #2

Hydroquinone	0.85 g	77 mM
Distilled H_2O	15 mL	
Subtotal	15 mL	

Solution #3

Silver lactate	0.11 g	5.5 mM
Distilled H_2O	15 mL	
Subtotal	15 mL	

Solution #4 Final Working Solution: First mix solutions #1 and #2; just before use add solution #3. The final working solution should be kept in the dark as much as possible, i.e., wrap working solution container and staining jars with aluminum foil.

RESULTS OF THE PROTOCOLS

Epithelial brush borders of kidney proximal tubules were labeled positively in oligonucleotide probe primed sections. All three oligonucleotide probes supported *in situ* transcription individually and as a mixture of the three. Cellular γ-GTP mRNA, demonstrated by IST, co-localized with γ-GTP protein detected by immunocytochemistry and with γ-GTP enzyme activity detected by enzymecytochemistry. All controls were negative.

These studies show that γ-GTP mRNA is also localized to the brush border of the cells rather than being scattered in the cytoplasm. These results suggest that γ-GTP mRNA is first transported to or near the functional location of the protein (in this case the apical brush border) where protein synthesis takes place [see Wang & De Bault (70)]. This is in contrast to another possibility where the mRNA is located centrally in the cell and the protein synthesis product is transported to its final functional site.

ACKNOWLEDGMENTS

We wish to thank Howard Doughty for his technical assistance in this project. This work was supported in part by NIH grant NS-18775 to LED.

REFERENCES

1. **Bains, M.A., R. Agarwal, J.H. Pringle, R.M. Hutchinson and I. Lauder.** 1993. Flow cytometric quantitation of sequence-specific mRNA in hemopoietic cell suspensions by primer-induced *in situ* (PRINS) fluorescent nucleotide labeling. Exp. Cell Res. *208*:321-326.
2. **Baltimore, D.** 1970. Viral RNA-dependent DNA polymerase. Nature (London) *226*:1209-1211.
3. **Bauman, J.G.J., J.A. Bayer and H. van Dekken.** 1990. Fluorescent *in-situ* hybridization to detect cellular RNA by flow cytometry and confocal microscopy. J. Microsc. *157*:73-81.
4. **Bayer, J.A. and J.G.J. Baumann.** 1990. Flow cytometric detection of γ-globulin mRNA in murine haemopoietic tissues using fluorescent *in situ* hybridization. Cytometry *11*:132-143.
5. **Blackburn, P.** 1979. Ribonuclease inhibitor from human placenta: Rapid purification and assay. J. Biol. Chem. *254*:12484-12487.
6. **Blackburn, P. and B.L. Jailkhani.** 1979. Ribonuclease inhibitor from human placenta: Interaction with derivatives of ribonuclease A. J. Biol. Chem. *254*:12488-12493.
7. **Blackburn, P., G. Wilson and S. Moore.** 1977. Ribonuclease inhibitor from human placenta. Purification and properties. J. Biol. Chem. *252*:5904-5910.
8. **Bradford, M.M.** 1976. A rapid and sensitive method for the quantitation of microgram quantities of protein utilizing the principle of protein-dye binding. Anal. Biochem. *72*:248-254.
9. **Caruthers, M.H.** 1985. Gene synthesis machine: DNA chemistry and its uses. Science *230*:281-285.
10. **Casperson, T., L. Zech, C. Johansson and E.J. Modest.** 1970. Identification of human chromosome by DNA-binding fluorescent agents. Chromosoma *30*:215-227.
11. **Chen, E.Y. and P.H. Seeburg.** 1985. Supercoil sequencing: A fast and simple method for sequencing plasmid DNA. DNA *4*:165-170.
12. **Chien, A., D.B. Edgar and J.M. Trela.** 1976. Deoxyribonucleic acid polymerase from the extreme thermophile Thermus acquaticus. J. Bacteriol. *127*:1550-1557.
13. **Cohen, S.N., A.C.Y. Chang, H. Boyer and R. Helling.** 1973. Construction of biologically functional bacterial plasmids *in vitro*. Proc. Natl. Acad. Sci. USA *70*:3240-3244.
14. **Cox, K.H., D.V. DeLeon, L.M. Angerer, and R.C. Angerer.** 1984. Detection of mRNAs in sea urchin embryos by *in situ* hybridization using asymmetric RNA probes. Dev. Biol. *101*:485-502.
15. **Craig, S. and C. Miller.** 1984. LR white resin and improved on-grid immunogold detection of vicilin, a pea seed storage protein. Cell Biol. Int. Rep. *8*:879-886.
16. **De Bault, L.E., E. Henriquz, M.N. Hart and P.A. Cancilla.** 1981. Cerebral microvessels and derived cells in tissue culture: II. Establishment, identification and preliminary characterization of an endothelial cell line. In Vitro *17*:480-494.
17. **De Bault, L.E., L.E. Kahn, S.P. Frommes and P.A. Cancilla.** 1979. Cerebral microvessels and derived cells in tissue culture: I. Isolation and preliminary characterization. In Vitro *15*:473-487.
18. **De Bault, L.E. and B.L. Wang.** 1994. Localization of mRNA by *in situ* transcription and immunogold-silver staining. Cell Vision *1*:67-70.
19. **De Bault, L.E., B.L. Wang and P. Grammas.** 1994. *In vitro* expression of γ-glutamyl transpeptidase (γ-GTP) and its mRNA in tube-forming structures in rat brain endothelial cells. J. Histochem. Cytochem. (In press).
20. **Eberwine, J., C. Spencer, K. Miyashiro, S. Mackler and R. Finnell.** 1992. Complementary DNA synthesis *in situ*: Methods and applications, p.80-100. *In* R. Wu (Ed.), Methods in Enzymology: Recombinant DNA, Vol. 216, Part G. Academic Press, San Diego.
21. **Eberwine, J., H. Yeh, K. Miyashiro, Y. Cao, S. Nair, R. Finnell, M. Zettel and P. Coleman.** 1992. Analysis of gene expression in single live neurons. Proc. Natl. Acad. Sci. USA *89*:3010.
22. **Eberwine, J.H., C. Spencer, D. Newell and A.R. Hoffman.** 1993. mRNA structure, *in situ*, as assessed by microscopic techniques. Micro. Res. Tech. *25*:19-28.
23. **Feinberg, A.P. and B. Vogelstein.** 1983. A technique for radiolabeling DNA restriction endonuclease fragments to high specific activity. Anal. Biochem. *132*:6-13.

24. **Finnell, R.H., G.D. Bennett, S.B. Karras and V.K. Mohl.** 1988. Common hierarchies of susceptibility to the induction of neural tube defects in mouse embryos by valproic acid and its 4-propyl-4-pentennoic acid metabolite. Teratology *38*:313.
25. **Finnell, R.H., S.P. Moon, L.C. Abbott, J.A. Golden and G.F. Chernoff.** 1986. Strain differences in heat-induced neural tube defects in mice. Teratology *33*:247.
26. **Gall, J.G. and M.L. Pardue.** 1969. Formation and detection of RNA-DNA hybrid molecules in cytological preparations. Proc. Natl. Acad. Sci. USA *63*:378-383.
27. **Glauert, A.M.** 1981. Practical Methods in Electron Microscopy: Fixation, Dehydration and Embedding of Biological Specimens. North-Holland Publishing Co., Amsterdam.
28. **Gosden, J., D. Hanratty, J. Starling, J. Fantes, A. Mitchell and D. Porteous.** 1991. Oligonucleotide-primed *in situ* DNA synthesis (PRINS): A method for chromosome mapping, banding, and investigation of sequence organization. Cytogenet. Cell Genet. *57*:100-104.
29. **Graber, M.B. and G.W. Kreutzberg.** 1985. Immuno gold staining (IGS) for electron microscopical demonstration of glial fibrillary acidic (GFA) protein in LR white embedded tissue. Histochemistry *83*:497-500.
30. **Gu, J.** 1994. Principles and applications of *in situ* PCR. Cell Vision *1*:8-19.
31. **Gubler, U. and B.J. Hoffman.** 1983. A simple and very efficient method for generating cDNA libraries. Gene *25*:263-269.
32. **Hacker, G.W., L. Grimelius, G. Danscher, G. Bernatzky, W. Muss, H. Adam and J. Thurner.** 1988. Silver acetate autometallography: An alternative enhancement technique for immunogold-silver staining (IGSS) and silver amplification of gold, silver, mercury and zinc in tissues. J. Histotechnol. *11*:213-221.
33. **Hacker, G.W., I. Zehbe, C. Hauser-Kronberger, J. Gu, A.H. Graf and O. Dietze.** 1994. *In situ* detection of DNA and mRNA sequences by immunogold-silver staining (IGSS). Cell Vision *1*:30-37.
34. **Harrison, P.R., D. Conkie, N. Affara and J. Paul.** 1974. *In situ* localization of globin messenger RNA formation. I. During mouse fetal liver development. J. Cell Biol. *63*:402-413.
35. **Hindkjar, J., J. Koch, J. Mogensen, S. Pedersen, H. Fischer, M. Nygaard, S. Junker, N. Gregersen, S. Kolvraa and L. Bolund.** 1991. *In situ* labelling of nucleic acids for gene mapping, diagnostics and functional cytogenetics, p.45–59. *In* J. Collins and A.J. Driesel (Eds.), Advances in Molecular Genetics, Vol. 4: Genome Analysis From Sequence to Function. Huthig Buch Verlag, Heidelberg.
36. **Holgate, C.S., P. Jackson, P.N. Cowen and C.C. Bird.** 1983. Immunogold-silver staining: New method of immunostaining with enhanced sensitivity. J. Histochem. Cytochem. *31*:938-944.
37. **Houts, G.E., M. Miyagi, C. Ellis, D. Beard and J.W. Beard.** 1979. Reverse transcriptase from avian myeloblastosis virus. J. Virol. *29*:517-522.
38. **Isaacson, S.H., D. Asher, C. Gajdusek and C.J. Gibbs, Jr.** 1994. Detection of RNA viruses in archival brain tissue by *in situ* RT-PCR amplification and labeled-probe hybridization. Cell Vision *1*:25-28.
39. **Janssen, K.** 1993. *In situ* hybridization and immunohistochemistry, p.14.3.1–14.3.7. *In* K. Janssen (Ed.), Current Protocols in Molecular Biology, 7th Edition. Breene Publishing Associates and John Wiley & Sons, USA.
40. **Janssen, K.** 1993. *In situ* hybridization and immunohistochemistry, p.14.2.1–14.2.6. *In* K. Janssen (Ed.), Current Protocols in Molecular Biology, 7th Edition. Breene Publishing Associates and John Wiley & Sons, USA.
41. **Janssen, K.** 1993. *In situ* hybridization and immunohistochemistry, p.14.1.4. *In* K. Janssen (Ed.), Current Protocols in Molecular Biology, 7th Edition. Breene Publishing Associates and John Wiley & Sons, USA.
42. **Kievits, T., J.G. Dauwerse, J. Wiegant, P. Devilee, M.H. Breuning, C.J. Cornelisse, G.J.B. van Ommen and P.L. Pearson.** 1990. Rapid subchromosomal localization of cosmids by nonradioactive *in situ* hybridization. Cytogenet. Cell Genet. *53*:134-136.
43. **Kleppe, K., E. Ohtsuka, R. Kleppe, I. Molineux and H.G. Khorana.** 1971. Studies on polynucleotides XCVI. Repair replication of short synthetic DNAs as catalyzed by DNA polymerases. J. Mol. Biol. *56*:341-361.
44. **Koch, J., H. Fischer, H. Askholm, J. Hindkjaer, S. Pedersen, S. Kolvraa and L. Bolund.** 1993. Identification of a supernumerary der(18) chromosome by a rational strategy for the cytogenetic typing of small marker chromosomes with chromosome-specific DNA probes. Clin. Genet. *43*:200-203.

45. **Koch, J., J. Hindkjaer, J. Mogensen, S. Kolvraa and L. Bolund.** 1991. An improved method for chromosome-specific labeling of alpha satellite DNA *in situ* by using denatured double-stranded DNA probes as primers in a primed *in situ* labeling (PRINS) procedure. Gene Anal. Tech. *8*:171-178.
46. **Koch, J., J. Mogensen, S. Pedersen, H. Fischer, J. Hindkjaer, S. Kolvraa and L. Bolund.** 1992. Fast one-step procedure for the detection of nucleic acids *in situ* by primer-induced sequence-specific labeling with fluorescein-12-dUTP. Cytogenet. Cell Genet. *60*:1-3.
47. **Koch, J.E., S. Kolvraa, K.B. Petersen, N. Gregersen and L. Bolund.** 1989. Oligonucleotide-priming methods for the chromosome-specific labelling of alpha satellite DNA *in situ*. Chromosoma *98*:259-265.
48. **Kogan, S.C., M. Doherty and J. Gitschier.** 1987. An improved method for prenatal diagnosis of genetic diseases by analysis of amplified DNA sequences. N. Engl. J. Med. *317*:985-990.
49. **Kohler, G. and C. Milstein.** 1975. Continuous cultures of fused cells secreting antibody of predefined specificity. Nature (London) *256*:495-497.
50. **Langer, P.R., A.A. Waldrop, and D.C. Ward.** 1981. Enzymatic synthesis of biotin-labeled polynucleotides: Novel nucleic acid affinity probes. Proc. Natl. Acad. Sci. USA *78*:6633-6637.
51. **Larsson, L.I.** 1989. Immunocytochemical detection systems, p.77-146. *In* L.I. Larsson (Ed.), Immunocytochemistry: Theory and Practice, 2nd Ed. CRC Press, Boca Raton.
52. **Larsson, L.I.** 1989. Fixation and tissue pretreatment, p.41-76. *In* L.I. Larsson (Ed.), Immunocytochemistry: Theory and Practice, 2nd Edition. CRC Press, Boca Raton.
53. **Lawrence, J.B. and R.H. Singer.** 1985. Quantitative analysis of *in situ* hybridization methods for the detection of actin gene expression. Nucleic Acids Res. *13*:1777-1799.
54. **Lawrence, J.B., K. Taneja and R.H. Singer.** 1989. Temporal resolution and sequential expression of muscle-specific genes revealed by *in situ* hybridization. Dev. Biol. *133*:235-246.
55. **Lichter, P., T. Cremer, C.J.C. Tang, P.C. Watkins, L. Manuelidis and D.C. Ward.** 1988. Rapid detection of human chromosome 21 aberrations by *in situ* hybridization. Proc. Natl. Acad. Sci. USA *85*:9664-9668.
56. **Lichter, P., C.J.C. Tang, K. Call, G. Hermanson, G.A. Evans, D. Housman and D.C. Ward.** 1990. High-resolution mapping of human chromosome 11 by *in situ* hybridization with cosmid clones. Science *247*:64-69.
57. **Longley, J., M.A. Merchant and B.M. Kacinski.** 1989. *In situ* transcription and detection of CD1a mRNA in epidermal cells: An alternative to standard *in situ* hybridization techniques. J. Invest. Dermatol. *93*:432-435.
58. **Luna, L.G.** 1986. Manual of Histologic Staining Methods of the Armed Forces Institute of Pathology, 3rd Edition. McGraw-Hill, New York.
59. **Mitchell, A., P. Jeppesen, D. Hanratty and J. Gosden.** 1992. The organization of repetitive DNA sequences on human chromosomes with respect to the kinetochore analysed using a combination of oligonucleotide primers and CREST anticentromere serum. Chromosoma *101*:333-341.
60. **Mogensen, J., S. Kolvraa, J. Hindkjaer, S. Petersen, J. Koch, M. Nygaard, T. Jensen, N. Gregersen, S. Junker and L. Bolund.** 1991. Nonradioactive, sequence-specific detection of RNA *in situ* by primed *in situ* labeling (PRINS). Exp. Cell Res. *196*:92-98.
61. **Nuovo, G.J., F. Gallery, P. MacConnell, J. Becker and W. Bloch. 1991.** An improved technique for the *in situ* detection of DNA after polymerase chain reaction amplification. Am. J. Pathol. *139*:1239-1244.
62. **Rentrop, M., B. Knapp, H. Winter and J. Schweizer.** 1986. Aminoalkylsilane-treated glass slides as support for *in situ* hybridization of keratin cDNAs to frozen tissue sections under varying fixation and pretreatment conditions. Histochem. J. *18*:271-276.
63. **Saiki, R.K., S. Scharf, F. Faloona, K.B. Mullis, G.T. Horn, H.A. Erlich and N. Arnheim.** 1985. Enzymatic amplification of beta-globin genomic sequences and restriction site analysis for diagnosis of sickle cell anemia. Science *230*:1350-1354.
64. **Sheehan, D.C. and B.B. Hrapchak.** 1980. Processing of tissue: Dehydrants, clearing agents, and embedding media, p.59-78. *In* D.C. Sheehan and B.B. Hrapchak (Eds.), Theory and Practice of Histotechnology, 2nd Edition. Battelle Press, Columbus.
65. **Shimomaye, E. and M. Salvato.** 1989. Use of avian myeloblastosis virus reverse transcriptase at high temperature for sequence analysis of highly structured RNA. Gene Anal. Tech. *6*:25-28.
66. **Tecott, L.H., J.D. Barchas and J.H. Eberwine.** 1988. *In situ* transcription: Specific synthesis of complementary DNA in fixed tissue sections. Science *240*:1661-1664.

67. **Temin, H.M. and S. Mizutani.** 1970. RNA-dependent DNA polymerase in virions of rous sarcoma virus. Nature (London) *226*:1211-1213.
68. **Trask, B.J.** 1991. Fluorescence *in situ* hybridization: Applications in cytogenetics and gene mapping. Trends Genet. *7*:149-154.
69. **Van Gelder, R.N., M.E. von Zastrow, A. Yool, W.C. Dement, J.D. Barchas and J.H. Eberwine.** 1990. Amplified RNA synthesized from limited quantities of heterogeneous cDNA. Proc. Natl. Acad. Sci. USA *87*:1663-1667.
70. **Wang, B.L. and L.E. De Bault.** 1994. Co-localization of γ-glutamyl transpeptidase and its mRNA in paraffin-embedded sections of rat kidney: Enzymecytochemistry, immunocytochemistry, *in situ* transcription and immunogold-silver staining systems. Cell Vision *1*:122-130.
71. **Ward, D.C., J. Menninger, J. Lieman, T. Desai, A. Banks, A. Boyle, P. Bray-Ward and T. Haaf.** 1994. Integration of the physical, genetic and cytogenetic maps of human chromosomes: Implications for the development of diagnostic DNA probes. Cell Vision *1*:61-66.
72. **Weih, F., A.F. Stewart and G. Schutz.** 1988. A novel and rapid method to generate single stranded DNA probes for genomic footprinting. Nucleic Acids Res. *16*:1628.
73. **Zehbe, I., G.W. Hacker, E. Rylander, J. Sallstrom and E. Wilander.** 1992. Detection of single HPV copies in SiHa cells by *in situ* polymerase chain reaction (*in situ* PCR) combined with immunoperoxidase and immunogold-silver staining (IGSS) techniques. Anticancer Res. *12*:2165-2168.

Address correspondence to Lawrence E. De Bault, Department of Pathology, P.O. Box 26901, BMSB 451, Oklahoma City, OK 73190, USA.

Sensitive Detection of DNA and mRNA Sequences by *In Situ* Hybridization and Immunogold-Silver Staining

Gerhard W. Hacker[1], Ingeborg Zehbe[2], Cornelia Hauser-Kronberger[1], Jiang Gu[3], Angelika Graf[4], Lars Grimelius and Otto Dietze[1]

[1]Institute of Pathological Anatomy, Immunohistochemistry and Biochemistry Unit, Salzburg General Hospital, and Medical Research Coordination Center, University of Salzburg, Salzburg, Austria; [2]University of Uppsala, Institute of Pathology, Uppsala, Sweden; [3]Deborah Research Institute, Browns Mills, NJ, USA; and [4]Salzburg General Hospital, Department of Gynecology and Obstetrics, Salzburg, Austria

SUMMARY

Recently, PCR in situ *hybridization (PISH) and* in situ *self-sustained sequence replication-based amplification (*in situ *3SR) methods have been described. They allow* in situ *detection of a single copy of DNA or mRNA. With these methods, the amplified nucleic acid sequence is detected by non-isotopic* in situ *hybridization (ISH) with biotin- or digoxigenin-labeled DNA or RNA probes to ensure a high specificity and to avoid false-positive stainings, sometimes resulting from direct incorporation of label. Conventional* in situ *hybridization is still frequently applied in cases where the copy number of nucleic acid sequences is above 20 copies per cell. The sensitivity and outcome of these methods partially depend on the detection system used. In this study, optimized ISH protocols were combined with immunogold-silver staining methods. The combinations described yielded a particulate, black staining of the structures containing the DNA or RNA sequence in question. The same detection method can also be applied to PISH or 3SR methods. IGSS-stained preparations can be transferred to the electron microscope and may therefore be used as tools for* in situ *analysis of specific nucleic acid sequences harboring, e.g., virus-infected cells, at the ultrastructural level. Other detection systems applicable for light microscopic ISH and PISH are also briefly discussed. They include fluorescent-, peroxidase- and alkaline phosphatase-based systems.*

INTRODUCTION

A silver amplification technique called autometallography (AMG) was introduced by Danscher (13,14), who was the first using this method to demonstrate colloidal gold in tissue sections (15). The AMG setup, which had been misleadingly

Table 1. Antibodies and Other Immunogold Reagents Used for Detection of *In Situ* Hybridization Products

Probe	Dilution	Code No.	Source
AuroProbe One anti-biotin, 1 nm gold	1/25	RPN 473	Amersham, London, UK
Goat anti-biotin, 1 nm gold	1/50	GAB1	BioCell, Cardiff, UK
Goat anti-biotin, 5 nm gold	1/50	EM.GAB5	BioCell
Sheep anti-digoxigenin, 1 nm gold	*****	*****	BioCell
Sheep anti-digoxigenin, 5 nm gold	*****	*****	BioCell

Note: The dilutions reported gave optimal visualization of the hybridization sites with no or very low levels of unspecific background staining.

called "physical development", was applied by Danscher for the detection of catalytic metals and enzyme histochemistry (13–17) and by Holgate and colleagues for immunocytochemistry (32). The latter method used silver-amplified colloidal gold particles adsorbed to antibodies, and was termed the immunogold-silver staining (IGSS) technique (32). During the last decade, IGSS has proven to be a reliable, highly sensitive and detection-efficient method for the demonstration of various substances in formalin- or Bouin's-fixed and paraffin-embedded tissues (e.g., 22–24,26,27,35,36,53). IGSS is not only more sensitive than most other immunocytochemical techniques, but also simple to perform and gives a delicate black staining of immunoreactive structures distinctly visible against the unstained background (22,32,36,53). Originally, IGSS was developed from an indirect immunogold-staining (IGS) method and the finding that metallic gold can be silver-amplified by AMG (13,14). Silver lactate was used as the ion source in AMG, forming shells of metallic silver around small gold particles. This process is catalyzed by the electron-donor hydroquinone in a low pH citrate buffer. Gold particles are increased in size by silver precipitation and conglomerate if sufficiently adjacent to each other, resulting in a greyish to black precipitate on relevant sites visible by both light and electron microscopy. A modified technique for the detection of gold and some other metals, silver acetate AMG, has been formally introduced (23). It is easy to perform, efficient, economical and less sensitive to light than other developers. It thus allows monitoring of the amplification process under light microscopic control.

For *in situ* detection of nucleic acids, a variety of *in situ* hybridization (ISH) methods are available. Non-isotopic probes possess great advantages over radioactively labeled probes, including the avoidance of hazardous radiation and a much shorter turnover period, i.e., from a few hours to two days instead of from several days to several months. They also give an improved resolution and are more economical to perform than radioisotope-labeled ISH (7–9,39,41,42,45,52,55). Biotin- or digoxigenin-labeled hybridization probes are used in daily research and histopathologic diagnosis to detect viral nucleic acids in infected cells, to investigate biosynthesis of peptides or proteins and to study genetic disease.

Table 2. Commercial Sources of Chemicals Used

Product	Code	Source
Silver acetate	85140	Fluka, Switzerland
Trisodiumcitrate-dihydrate	6448	Merck, Germany
Citric acid	244	Merck
Hydroquinone	53960	Fluka
Tween 80	56023	BDH Chemicals, UK
Agefix	-	Agfa-Gevaert, Germany

Optimized protocols for DNA-ISH with biotin-labeled probes and IGSS detection have been described (25–27). Most recently, applications of AMG for polymerase chain reaction (PCR) *in situ* hybridization (PISH), or *in situ* self-sustained sequence replication-based amplification (*in situ* 3SR) have been suggested (58–61). These methods allow detection of single or very low copy numbers of specific DNA or mRNA sequences at the cellular level using IGSS and silver acetate AMG for the first time. Using these techniques in pre-embedding methods on formalin- or paraformaldehyde-fixed cells followed by osmium tetroxide post-fixation and resin-embedding, AMG of nucleic acid staining at the electron microscope (EM) level becomes possible (58; and Dr. Wolfgang H. Muss, Salzburg, Austria, personal communication).

We present here optimized DNA and mRNA *in situ* hybridization protocols and a protocol for the detection of hybridization sites with IGSS and silver acetate AMG. Other detection methods frequently applied are also discussed.

DNA-DNA *IN SITU* HYBRIDIZATION

For DNA-DNA *in situ* hybridization, our test system utilized tissue samples taken after surgery for human genital condylomata acuminata, cervical cancer, cytomegalovirus-infected lung and Epstein Barr virus containing Burkitt's lymphoma (24,25). The specimens were fixed in 4% phosphate-buffered formaldehyde for 12–16 h at room temperature, washed in 20 mM phosphate-buffered saline (PBS, pH 7.2), dehydrated in increasing concentrations of alcohols, cleared in xylene and routinely embedded in paraffin. In addition, cytospins of cell cultures containing the virus in question were used. Sections (5 μm) and cell preparations were mounted on silanized [aminopropyl-triethoxysilane (APES)-coated] glass slides (24,43). A range of commercially available biotinylated DNA probes specific for viral genomes was applied, and a highly stringent hybridization protocol published previously was used (25; see Protocol 1).

Briefly, deparaffinized sections or cytological preparations were immersed in PBS containing Triton® X-100, digested proteolytically with 0.1% proteinase K, washed in 2× standard saline citrate (SSC), transferred to isopropanol and air-dried.

After prehybridization, one of the two following biotinylated probe systems was used: 1) Ready-made probe solutions from Enzo Diagnostics (New York, NY, USA) or 2) nick-translated cDNA probes to different human papillomavirus (HPV) subtypes, dissolved in 50% formamide and 10% dextran sulfate in 2× SSC. Twenty microliters of ready-to-use probes or 20 ng of nick-translated probe were put onto the section, covered by a 22- × 22-mm coverslip, placed on a 92°C heating block for denaturation and further incubated at 37°C. Coverslips were then removed by soaking in 4× SSC. They were then washed with 4× SSC, 2× SSC, 0.1× SSC, 0.05× SSC and distilled water.

The hybridization sites were detected using a direct IGSS method and 1-nm gold-labeled goat anti-biotin antibodies with subsequent silver acetate AMG (see Protocol 3). IGSS-stained sections or cells were counterstained with eosin, Nuclear Fast Red or Neutral Red and coverslipped using DPX® (BDH Chemicals, Poole, UK) as mounting medium.

Negative controls to test the specificity of the technique included omission or replacement of the probe by hybridization mixture without specific probe or containing an unrelated probe and replacement of immunogold reagent by antibody diluent or normal sera. To test positive staining, various sections known to contain the nucleotide sequence in question were used.

For comparison between IGSS detection and other detection methods, hybridization sites were also detected using a streptavidin-biotin-peroxidase complex with 3-amino-9-ethylcarbazole (AEC) or 3,3′ diaminobenzidine tetrahydrochloride (DAB) for the visualization of peroxidase. These chromogens gave an alcohol-soluble reddish or permanent brown precipitate, respectively. After counterstaining with Mayer's hematoxylin, sections were mounted with Glyceromount® for AEC (Dako, Glostrup, Denmark) or DPX for DAB. In addition, an alkaline-phosphatase-based system using NBT/BCIP development was used.

RNA-RNA *IN SITU* HYBRIDIZATION

Atrial natriuretic peptide (ANP) in rat atria was used as the model for RNA-RNA *in situ* hybridization. Freshly collected hearts of adult rats were fixed in 4% paraformaldehyde for 3 h, processed and embedded in paraffin. Five-micron-thick sections were cut and mounted on silanized glass slides (43). A biotin-labeled riboprobe was used. Briefly, biotinylated single-stranded RNA probes complementary to mRNA encoding ANP were prepared using the Riboprobe system (Promega, Madison, WI, USA).

The procedures for probe preparation were those described previously (20,21). The ANP cDNA was a 0.8 kb *Hin*dIII-*Eco*RI fragment corresponding to rat α-ANP 1–28. The fragments were subcloned in the transcription Vector pSP65 (Promega) 3′ to 5′ vis-a-vis SP6 promoter. The transcription reaction was carried out in the volume of 50 μL in a buffer of 40 mM Tris, pH 7.5, 7 mM $MgCl_2$, 2 mM spermidine and 1 mM dithiothreitol (DTT). Both antisense and sense (using vector pSP64; Promega) probes were produced. The latter was used in the negative control. All stock solutions were made in diethylpyrocarbonate (DEPC)-treated water, autoclaved and filter-sterilized. One gram of each plasma subclone, linearized with appropriate restriction enzymes, was transcribed per reaction in the presence of 500 μg ATP, CTP, GTP and allylamine UTP (Life Technologies, Gaithersburg, MD, USA). Fifteen units of SP6 RNA polymerase were added per reaction and the mixture was incubated at 42°C for 30 min, after which an additional 15 units of SP6 polymerase were added and incubation was continued for an additional 30 min. Ten units of RNase-free DNase (RQ1; Promega) and the RNase inhibitor RNasin® (Promega) were added to the reaction, which was then incubated for 10 min at 37°C. Yeast tRNA carrier (50 μg) was added following the addition of 0.5 M salt, the mixture was extracted once with phenyl/chloroform, and once with chloroform/isoamyl alcohol. After which, ethanol was precipitated 3 times with resuspension solution consisting of 0.2% sodium dodecyl sulfate (SDS), 2 mM EDTA and 0.3 mM NH_4Ac. After the third precipitation, the pellets were resuspended in 10 mM Tris, 1 mM DTT and an aliquot was precipitated with trichloroacetic acid.

The primary amine of the incorporated allylamine UTP was biotinylated following repeat ethanol precipitation of the reaction mixture and resuspension in 50 μL of 0.1 M sodium borate buffer. Ten microliters (10 mg/mL) of CAB-NHS ester (Life Technologies) were added to the reaction mixture and incubated at 20°C for 2 h. The reaction mixture was made in 2 M NH_4Ac, ethanol-precipitated and resuspended in 10 mM Tris, 1 mM DTT. Aliquots of each mixture were run on formaldehyde gels. The RNA was then transferred to nitrocellulose filter paper and reacted to avidin-biotin-peroxidase complex (Vector Laboratories, Burlingame, CA) in PBS for 60 min. Peroxidase was visualized with DAB and H_2O_2. A distinct brown band was revealed on each gel at the correct position to verify that the probes had the correct size and had biotin effectively labeled to them.

The whole procedure was performed under sterile conditions to avoid contamination by RNase. RNA-RNA *in situ* hybridization was based on the technique previously employed by the authors (20,21). Briefly, after treating the tissue sections with proteinase K, the positive charge of the tissue was neutralized by immersing the slides in 0.1 M triethylamine, pH 8.0, with freshly added acetic anhydride (0.25%) and stirred gently for 10 min. The slides were rinsed again in 2× SSC for 1 min and re-rinsed in PBS for 1 min, with agitation, and then immersed in 0.1 M Tris-HCl, pH 7.0, containing 0.1 M glycine for 30 min, and rinsed in two changes of 2× SSC, 1 min each and then set in 2× SSC until ready to hybridize (up to several hours). Approximately 5 min before hybridization, the slides were transferred to 50% formamide in 2× SSC at 55°C.

The RNA hybridization solution consisted of 10 μL formamide with 20% dextran sulfate, 2 μL 20× SSC containing 100 mM DTT, 2 μL *Escherichia coli* tRNA (0.1 g/mL), 2 μL herring sperm DNA (10 mg/mL), 2 μL BSA (bovine serum albumin, nuclease-free, 20 mg/mL), and 2 μL labeled RNA probe. A pre-hybridization was performed using the hybridization mixture without the specific probe at 42°C for 30 min in a humid chamber. The above probe mixture was heated at 90°C for 10 min and transferred to 55°C before use. The excess liquid around the tissue section was blotted away with filter paper and 20 μL of the probe mixture was applied. The samples were coverslipped and incubated in a humidified chamber for 3 h at 50°C.

The slides were then washed with constant agitation in 3 changes of 50% formamide in 2× SSC at 52°C for a total of 30 min, and in 4 changes of 2× SSC at room temperature, 1 min each. Thirty microliters of RNase solution (100 μg/mL RNase A, 1 μg/mL RNase T1) were applied on each slide and incubated for 30 min at 37°C. After washing twice, the slides were rinsed again in 50% formamide in 2× SSC for 5 min at 52°C with constant agitation and twice in 2× SSC at room temperature for a total of 15 min. They were then ready for the IGSS detection followed by the same protocol as described for DNA ISH. For comparison, the labeling biotin was also detected by a streptavidin-biotin-peroxidase complex system followed by DAB as the chromogen. The efficiency and specificity of IGSS detection were examined side by side with the enzyme detection system.

RESULTS

All IGSS-ISH procedures described above gave positive results with distinct grey to black granular deposits at the hybridization sites (Figures 1–3). Detection of papillomaviruses in oral or cervical lesions was confined to nuclei of HPV harboring keratinocytes and/or koilocytes (Figures 1 and 2), whereas cytomegalovirus (CMV) was distinctly visualized in infected "owl-eye" lung cells (Figure 2). Appropriate reactions were also seen in tissue sections or cell cultures containing adenovirus 5, hepatitis B virus, Epstein Barr virus or herpes simplex virus.

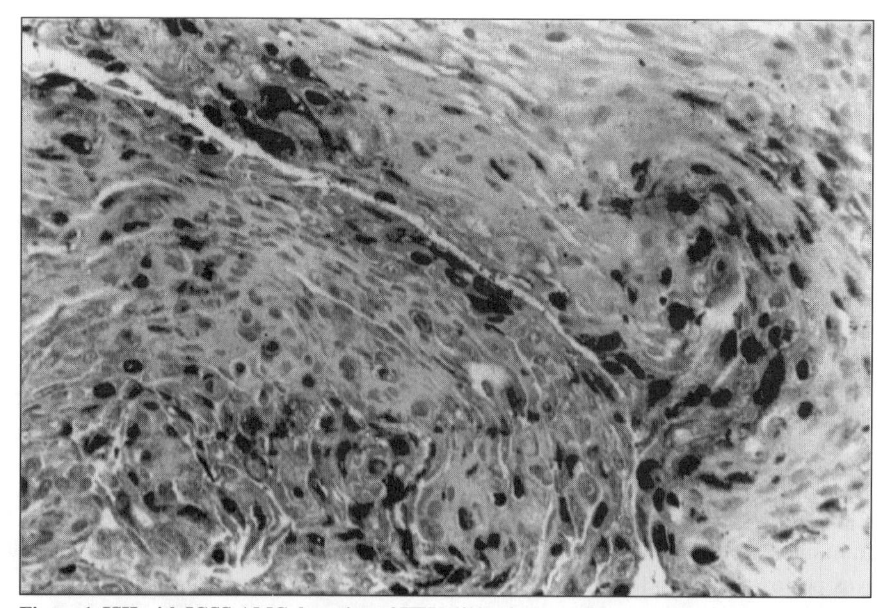

Figure 1. ISH with IGSS-AMG detection of HPV 6/11 using a double-stranded, biotinylated DNA-probe. The section shows viral infection of a medium-grade cervical dysplasia. Virus-harboring cells show distinct black staining of the nuclei and sometimes also in the cytoplasm, possibly representing virus-packaging sites. Formalin-fixed, 5-μm-thick paraffin section, H&E counterstained. Magnification, ×220.

The DNA ISH of formalin-fixed paraffin sections pretreated with proteinase K showed a higher signal-to-noise ratio and a higher sensitivity compared to proteolysis with trypsin. Duration of the predigestion period and concentration of proteinase K had to be optimized according to the strength of tissue fixation and the size and thickness of the section. The use of acetic anhydride in triethanolamine and incubating with formamide and dextran sulfate in 2× SSC did not influence the staining appearance in this particular setup. Air-drying of the sections after dehydration in graded ethanols before pre-hybridization treatments or hybridization gave a stronger staining with very low background. Treatment with Lugol's iodine after the stringent washes following the actual hybridization reaction gave a better signal-to-noise ratio than when this oxidizing step was omitted.

The RNA-RNA hybridization system gave strong and specific detection of ANP mRNA in rat heart atrial cardiocytes (Figure 3) and were absent from the ventricles. The positivity was confined to the cytoplasm, particularly the paranuclear areas. The specificity of the reaction was confirmed by the results of the appropriate controls. In comparison to the enzyme detection, the IGSS gave a more distinct black staining without background, with a signal-to-noise ratio better than that obtained with the enzyme detection system.

The optimal working protocols for DNA-DNA and RNA-RNA hybridization, as well as IGSS detection of the hybridization sites, are given in Protocols 1–3.

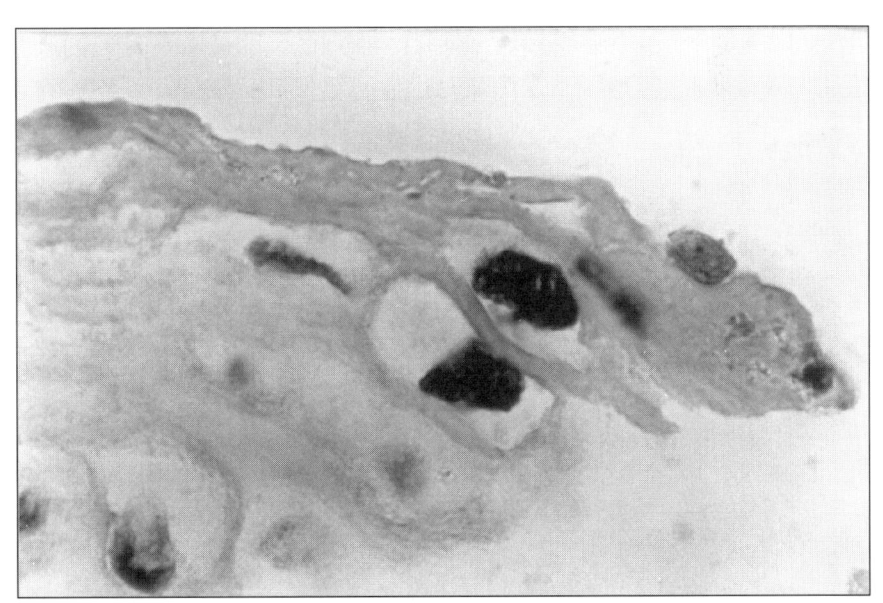

Figure 2. ISH with IGSS-AMG detection of HPV 16/18 from a co-infection of different areas of condylomata acuminata with HPV 6/11 and HPV 16/18. HPV 16/18-harboring nuclei are distinctly stained in black color. High-power photomicrograph of a formalin-fixed, 5-µm-thick paraffin section, H&E counterstained. Magnification, ×720.

RECOMMENDED PROTOCOLS

Protocol 1: *In Situ* DNA Hybridization (24,25)

This protocol is still under development and is only given as a guideline.

1. Thick (5–10 µm) cryostat sections are cut, mounted on APES-coated glass slides (43), dehydrated in graded ascending alcohols, put into xylene for 30 min and rehydrated in graded descending alcohols. (APES coating of glass slides: Clean slides with acetone, let them dry and incubate with 2% 3-(triethoxysilyl)-propylamine [APES; Merck, Darmstadt, Germany] dissolved in acetone for 5 min. Then wash slides in acetone and distilled water and dry them again).

2. Post-fix sections in neutral 5% phosphate-buffered formaldehyde or 2% buffered paraformaldehyde (10 min).

3. Wash sections in 20 mM PBS pH 7.2 (3× 2 min).

4. Soak in 0.3% Triton X-100 (15 min) to permeabilize sections.

5. Proteolytic treatment: Lightly digest the tissue sections with 0.1% proteinase K in PBS for about 15 min (necessity, time and concentration depend on the strength of pre-fixation of the tissue and the size of the section) at 37°C using a water bath.

6. Wash in PBS or 2× SSC (3× 2 min).

7. Wash in distilled water (2 min), immerse in 50%, 70% and 98% isopropanol (1 min each) and air-dry at room temperature.

8. Optional—Prehybridization: Incubate with 50% deionized formamide and 10% dextran sulfate in 2× SSC, at 50°C for 5 min. Drain off excess. Special care must be taken that sections never dry from the hybridization step on.

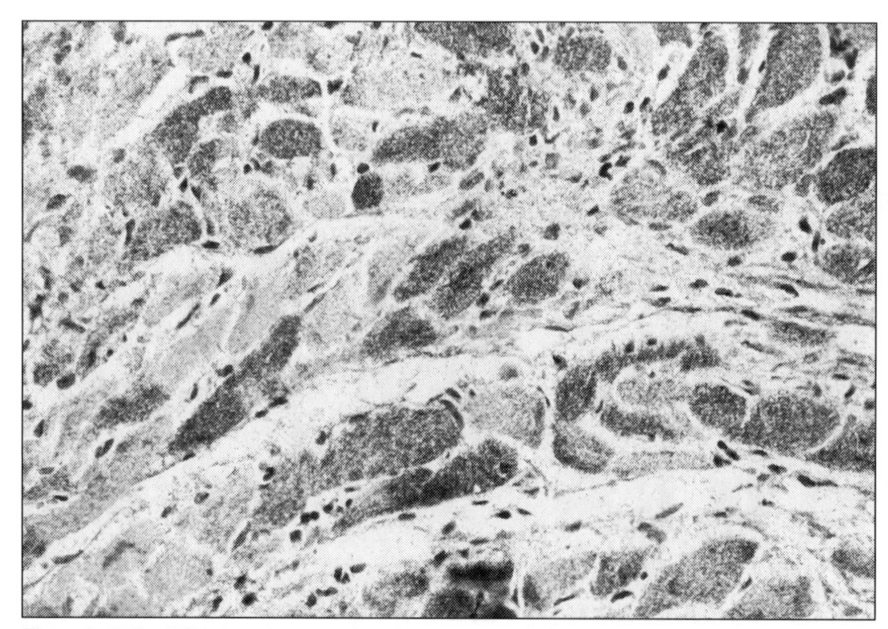

Figure 3. ANP-mRNA detection. *(A color reproduction of this figure appears on p. 143.)*

9. Put a small drop of probe mixture (about 20 μL of ready-to-use probes or 20 ng of nick-translated probe) onto the section, which is then covered with a 22- × 22-mm coverslip. In this setup, biotinylated or digoxigenin-labeled cDNA probes have been successfully used.

10. Place the slides on a 92°C heating block and incubate for 4–10 min.

11. Move the slides into a 37°C oven and incubate for further 60–120 min.

12. Remove coverslips by soaking with 4× SSC.

13. Wash under stringent conditions in 4× SSC, 2× SSC, 0.1× SSC, 0.05× SSC, and distilled water (each > 5 min) at room temperature or (better) at 37°C.

14. Immerse in Lugol's iodine (1% iodine in 2% potassium iodide) (5 min) and briefly rinse in distilled water.

15. Treat with 2.5% aqueous sodium thiosulfate until sections become colorless and wash in distilled water (3× 2 min).

16. Immerse in TBS-gelatin (Tris-buffered saline pH 7.6 containing 0.1% cold water fish gelatin) (2× 3 min).

17. Incubate with gold-adsorbed anti-biotin (Amersham, London, UK) or anti-digoxigenin (Boehringer Mannheim, Mannheim, Germany; Aurion, Wageningen, NL; or BioCell, Cardiff, UK) antibodies for at least 60 min at room temperature or overnight at 4°C. Optimum dilution is between 1/25 and 1/50. Antibody diluent is TBS-gelatin containing 0.8% BSA.

18. Wash in TBS-gelatin (3× 3 min).

19. Post-fix in 2% glutaraldehyde in PBS pH 7.2 (2 min).

20. Apply silver acetate autometallography as in Protocol 3. Do not counterstain.

Note: Preparation of 20× SSC: 175.32 g NaCl, 88.23 g Na-citrate in 1 L H_2O; adjust to pH 7.0 with HCl or citric acid; premixed concentrate is available from Sigma (No. S-6639). Steps 14 and 15 should be avoided if the protocol is used for detection of *in situ* PCR.

Protocol 2: *In Situ* mRNA Hybridization (20,21,25)

Special care must be taken not to touch slides or solutions with bare fingers or other possibly RNase-contaminated materials. Wear gloves. Use RNase-free (DEPC-treated) and sterilized water for all washes and solutions.

1. Cut paraffin sections and mount them on silanized (APES-coated) slides as described in Protocol 1. Bake sections at 70°C for 25 min, deparaffinize with xylene, rehydrate with graded alcohols and take to TBS.

2. Pre-digest sections with proteinase K, e.g., with 0.3 mg proteinase K per milliliter buffer pH 7.2 at 37°C for 15 min. (**Note:** This is a guideline only, as the duration and concentration necessary are to be assessed for the particular setup and batch of proteinase K used.)

3. Wash in double-distilled water for 5 min.

4. Deactivate remnants of proteinase K on hot plate or in a thermocycler at 95°C for 1 min.

5. Treat with DNase for 25 min at room temperature, if desired.

6. Neutralize the positive charge of the slides by immersing them in 0.1 M triethylamine, pH 8.0, with freshly added acetic anhydride (0.25%) and stir gently for 10

min. This step is optional depending on the amount of background staining obtained.

7. Rinse the slides in 2× SSC for 1 min and in PBS for 1 min with agitation.

8. Immerse in 0.1 M Tris-HCl, pH 7.0, containing 0.1 M glycine, for 30 min.

9. Rinse the slides in two changes of 2× SSC, 1 min each, and then leave in 2× SSC until hybridization (up to a few hours).

10. Prepare pre-hybridization mixture: 250 μL dextran sulfate, 149 μL 20× SSC, 120 μL 1 mM EDTA, 33 μL herring sperm DNA, and 448 μL double-distilled water, to reach 1000 μL final pre-hybridization mix.

11. Apply pre-hybridization mixture at 42°C for 30 min in a humid chamber, and then wash in 2× SSC.

12. Hybridization mixture for oligoprobe: 250 μL dextran sulfate, 120 μL 1 mM EDTA, 100 μL 20× SSC, 450 μL deionized formamide, 33 μL herring sperm DNA, and 47 μL double-distilled water. Biotinylated probe stock solution is 30 ng/μL. 1.4 μL of this probe stock solution are dissolved in 400 μL hybridization mix, giving a final concentration of 0.1 ng/μL.

Hybridization mixture for riboprobe: 10 μL formamide with 20% dextran sulfate, 2 μL 20× SSC containing 100 mM DTT, 2 μL *E. coli* tRNA, 2 μL herring sperm DNA (10 mg/mL), 2 μL BSA (nuclease-free, 20 mg/mL) and 2 μL labeled RNA probe.

13. Denature DNA before applying the probe by heating the sections for 5–10 min on a 92°C heating block.

14. Transfer the sections to a 42°C warm humid chamber and incubate overnight.

15. Wash in 5× SSC for 5 min.

16. Wash in 2× SSC for 5 min.

17. Proceed to detection with IGSS and silver acetate autometallography, or streptavidin-biotin-peroxidase or streptavidin-alkaline phosphatase systems.

Protocol 3: Highly Sensitive IGSS Method for Detection of Biotin- or Digoxigenin-Labeled Hybridization Sites

A. Immunocytochemical Detection by IGSS (25,27)

1. Following the hybridization and washing under stringent conditions, the sections or cytospins are immersed in Lugol's iodine (1% iodine in 2% potassium iodide; ready-made from Merck No. 9261, Darmstadt, Germany) (5 min).

2. Rinse briefly in distilled water.

3. Treat with 2.5% aqueous sodium thiosulfate until sections become colorless (up to 30 s).

4. Wash in distilled water (2 min).

5. Immerse in TBS-gelatin (TBS pH 7.6 containing 0.1% cold water fish gelatin) (10 min). In some cases, superior results are obtained if the buffer in this step also contains 0.1% Triton X-100 or Tween®-80, and 2.5% NaCl.

6. Apply normal serum of the species providing the secondary antibody (1/10 in TBS-gelatin) (5 min) and drain off.

7. Incubate with 1 and/or 5 nm gold-labeled antibody against biotin or digoxigenin (overnight at 4°C). The dilution should be optimized carefully. Suggested

antibody diluent is 0.1 M TBS or PBS (pH 7.6) containing 0.1% BSA and 0.1% sodium azide.

8. Wash in TBS-gelatin (3× 3 min).

9. Apply normal serum 1/10 as in step 8.

10. Incubate with 5 nm gold-adsorbed second layer antibodies directed against immunoglobulin of the species providing the gold-labeled primary antibody (60 min at room temperature). Optimum dilution is usually between 1/25 and 1/100 and should be determined by titration. The use of the same antibody diluent as in step 7, or TBS-gelatin, is suggested.

11. Wash in TBS-gelatin (3× 3 min).

12. Post-fix in 2% glutaraldehyde in PBS pH 7.2 (2 min).

13. Rinse briefly 5 times in distilled water (about 30 s each), followed by 3 washes (3 min each) in double-distilled water.

16. Perform silver acetate autometallography.

Note: Steps 1–4 are not used when *in situ* PCR has been carried out prior to IGSS detection! Steps 9–11 are optional and may be applied to further enhance the sensitivity of the detection system. The gold-labeled anti-biotin antibodies used for the detection of biotinylated probes may also be replaced by streptavidin-gold (with omission of steps 9–11); however, this may result in decreased sensitivity.

B. Silver Acetate Autometallography (23)

1. Solutions A and B should be freshly prepared.

Solution A: Dissolve 80 mg silver acetate (code 85140; Fluka, Buchs, Switzerland) in 40 mL of glass double-distilled water. Silver acetate crystals can be dissolved by continuous stirring for about 15 min.

2. Citrate buffer: Dissolve 23.5 g of trisodium citrate dihydrate and 25.5 g citric acid monohydrate in 850 mL of deionized or distilled water. This buffer can be kept at 4°C for at least 2–3 weeks. Adjust to pH 3.8 with aqueous citric acid solution before use.

3. Solution B: Dissolve 200 mg hydroquinone in 40 mL citrate buffer.

4. Just before use, mix solutions A and B.

5. Silver amplification: Immerse the slides vertically in the mixture of solutions A and B in a glass container (preferably a Schiefferdecker container with about 80 mL volume, in which up to 19 slides can be developed at once) and cover the setup with a dark box. Staining intensity may be checked and optimized under a light microscope adjusted to low power illumination during the amplification process.

6. When optimal labeling is achieved, the amplification process can be stopped with a photographic fixer (e.g., Agefix® [Agfa Gevaert, Leverkusen, Germany] diluted 1:20 with water). This fixing solution can be reused. Treat the slides for about 1 min. Alternatively, a 2.5% aqueous solution of sodium thiosulfate may be used.

7. Rinse the slides carefully in tap water for at least 3 min. Silver-amplified sections can now be counterstained with hematoxylin and eosin, Nuclear Fast Red, Methyl Green or Neutral Red, dehydrated and mounted in DPX.

Note: For every kind of silver amplification (AMG), it is advisable to use thoroughly clean glassware. To yield optimally clean glass, rinsing with Farmer's reducer solution followed by thorough washing in distilled water is recommended

(17). In order to decrease the intensity of staining in sections which have been overdeveloped or which have a high level of unspecific background, a modified Farmer's reducer can be applied (53).

DISCUSSION

ISH is a well-established technology for demonstrating cellular distribution of specific nucleotide sequences by using labeled double-stranded DNA probes or single-stranded DNA or RNA probes complementary to the sequence under investigation (3–5,7,8,12,19,31,33,37,48,55). ISH with biotin- or digoxigenin-labeled probes is a powerful technique for the detection of viral DNA and gene expression (7,20,41,42,45,52,55). However, the detecting sensitivity is low when compared to PCR-based or related systems (40,46,47,57–61) (Figures 4 and 5).

Enzyme-labeled streptavidin-biotin methods for ISH have been proposed (e.g., 5,6,11,44,48), and various protocols for the detection of hybridization sites with fluorescent or enzyme-based methods have been suggested (e.g., 1,2,10,30,34, 38,48,49,52). The detection sensitivity of ISH depends to a high degree on the system used for detection of the probe (31,45). Some authors believe that alkaline

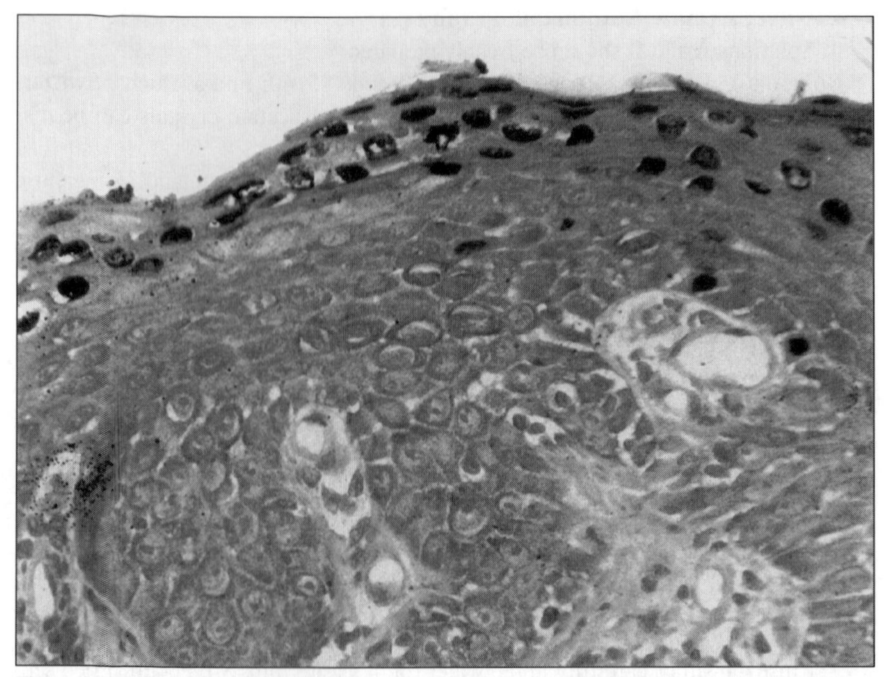

Figure 4. PISH of HPV in condylomata acuminata. A primer pair outlining a consensus sequence appearing in most HPV subtypes was used (60). *In situ* PCR was followed by ISH with digoxigenin as the reporter molecule and IGSS-AMG. This procedure resulted in a distinct black signal seen in virus-harboring nuclei. Following the PISH procedure, the section was additionally treated with the streptavidin-biotin-peroxidase complex (S-ABC) technique and "KL-1" monoclonal antibodies to cytokeratins, which gave a distinct brown stain of the cytoplasmic intermediate filaments in epithelial cells. Formalin-fixed, 5-μm-thick paraffin section, hematoxylin counterstained. Magnification, ×290.

phosphatase is the most suitable enzyme for non-radioisotopic ISH detection (9,45). However, in our experience alkaline phosphatase does not always perform better than peroxidase. It has been calculated that the detection limit of biotinylated probes in combination with peroxidase as the label may be as low as about 20 copies per cell (56,59). Enzymatic or fluorescent labels avoid the use of hazardous radioisotopes, and the probes can be stored and handled easily. However, these detection methods also have disadvantages in comparison to the particulate IGSS methods. FITC- and alkaline phosphatase-based systems are generally regarded as non-permanent stainings: fluorescent labels tend to fade during UV light exposure. For demonstrating peroxidase, potentially carcinogenic substances such as DAB, 4-chloro-1-naphthole or AEC are used. The latter two chromogens are alcohol-soluble, which is only compatible with aqueous mounting media. Peroxidase-DAB-based techniques, in our experience, appear to give less spatial resolution than IGSS (25).

As first shown by Varndell et al. (55) and subsequently by other groups (19,25,28,29,39,41,42), gold-adsorbed macromolecules can substitute alkaline phosphatase or other enzymes used for immunolabeling ISH. Roth et al. (50,51)

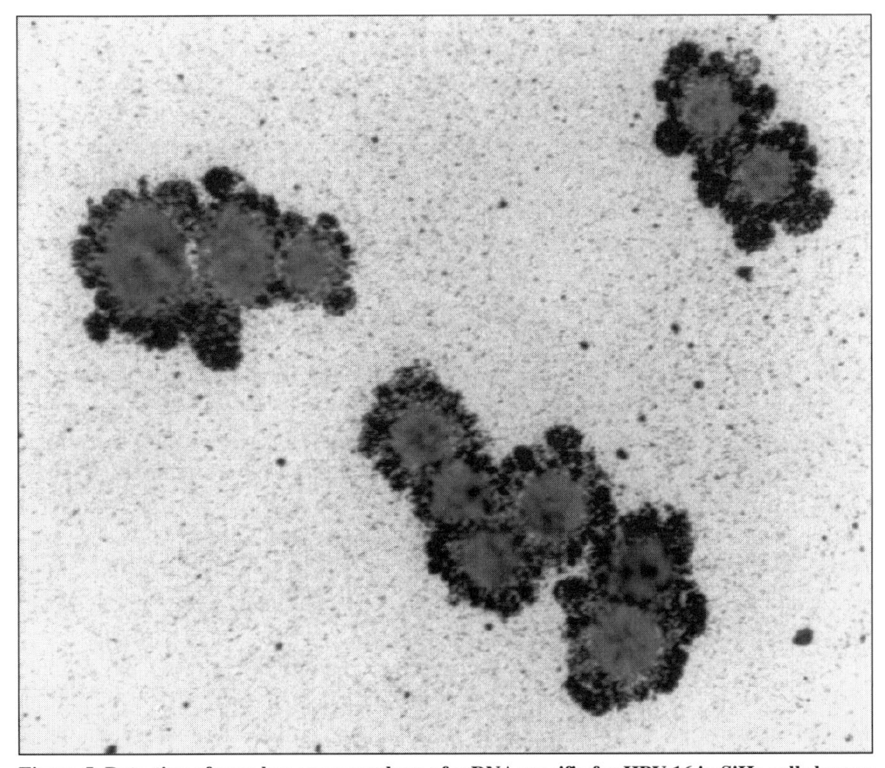

Figure 5. Detection of very low copy numbers of mRNA specific for HPV-16 in SiHa cells by our technique of *in situ* self-sustained sequence replication-based amplification (*in situ* 3SR) reaction with ISH-IGSS-AMG detection (61). The amplified mRNA containing sites are distinctly visualized as a black stain in the cytoplasm. A digoxigenin-labeled hybridization probe was used. Paraformaldehyde-fixed SiHa cell culture, counterstained with Nuclear Fast Red. Magnification, ×600.

have reported very promising results using a gold-labeled horseradish peroxidase antibody together with silver amplification. In conventional IGSS detection, some authors (18,25,39,45,55) used streptavidin-gold and AMG directly after ISH; others (28,29,41,42) used unlabeled anti-biotin antibodies and the indirect streptavidin-gold-silver technique (11). A method comparable to our protocol was given by Giaid et al. (19); however, these authors utilized larger gold particles. One nanometer gold is now available and has proved to be very efficient (e.g., 24–27). A recently published study employing several different IGSS protocols has shown that the various IGSS methods gave different staining intensities and therefore had different detection efficiencies (24). The outcome of IGSS procedures also strongly depends on the system used for gold amplification. Our technique of silver acetate AMG was designed to give maximum detecting sensitivity and efficiency with minimal light sensitivity and background (23). Commercially manufactured "easy-to-use" developers in most cases do not satisfy our requirements of giving a fast, specific, relatively light-insensitive, detection-efficient AMG amplification. Different developers and the use of different silver salts have recently been compared in detail at both light microscopic and EM levels (27,54). So far, we have found only one commercial kit (code No. SEKL15; BioCell) to be of real use—fulfilling all our criteria.

In theory, indirect staining methods should give superior results to the direct methods. However, new immunogold-reagents, i.e. gold-adsorbed anti-biotin or anti-digoxigenin antibodies, provide ways to overcome this problem by giving a very high signal-to-noise ratio and being more economical in comparison to other methods (24–27,60,61). The use of these reagents is described in Protocol 3. The smaller gold particles, 1 nm in diameter, have resulted in a very discrete staining appearance when compared to larger gold particles. Also, the small gold size may help to improve penetration of the immunogold reagent into the nucleus for DNA detection. Interestingly, streptavidin-gold, in our hands, gave a lower staining intensity than direct or indirect IGSS methods. This might be attributed to the steric hindrance of molecules (24). The very best labeling intensity in the present study was obtained when a combination of a direct with an indirect labeled IGSS procedure was performed sequentially on the same preparation. After the hybridization washes, sections were incubated with gold-labeled anti-biotin- or anti-digoxigenin antibodies, followed by a gold-labeled antibody recognizing immunoglobulin of the species in which the primary antibody was produced. Using this technique, several cases of condylomata acuminata that were negative for HPV with conventional peroxidase- or alkaline-phosphatase-labeled systems became positive. This phenomenon deserves further investigation.

Compared to other detection systems, IGSS allows the use of conventional counterstains in full strength, such as haematoxylin and eosin (H&E), Neutral Red, Methyl Green or Nuclear Fast Red, which greatly improves morphological assessment. In most other detection systems, only a very light nuclear counterstain can be used. The dark grey to black silver AMG stain can be identified easily at low magnification, which is of particular help for accurate histopathological diagnosis (22–24,32,53). The high signal-to-noise ratio of IGSS is especially suitable for computer image analysis (25). The high level of unspecific background staining described by Giaid et al. (19), who used a commercially available silver intensification kit, was not found in our study as our stainings were optimized by light microscopical control using silver acetate AMG. As discussed earlier (23), silver acetate AMG is less sensitive to light than silver lactate developers (13–17,27,35). The detection protocol described in the present study has also been applied successfully in lectin histochemistry (unpublished observations), and to PCR *in situ* hybridization and *in situ* 3SR (57,60,61) (Figures 4 and 5). Using the IGSS-ISH methods described here, it is also possible to further process the specimens for EM examination, i.e., as a pre-embedding ISH method using a "pop-off" technique (58).

Lugol's iodine oxidation with subsequent decolorization using sodium thiosulfate significantly increases the detection efficiency of IGSS methods (32,53). For ISH, these steps should be carried out after the hybridization step (24), a finding confirming the results of Henke et al. (29). For PCR-ISH (PISH), we found that Lugol's iodine treatment should be avoided, as it creates holes in the cell and nuclear membranes and therefore possibly promotes diffusion of the PCR product (57). It was found that the addition of 0.1% cold water fish gelatin to the washing buffers considerably increases the overall performance of IGSS, possibly due to its function as a protective colloid in AMG (25–27). In IGSS-detected PISH, the gelatin in combination with glutardialdehyde post-fixation may further help to prevent leakage of the amplified DNA signal from the nucleus.

In conclusion, detection of DNA-DNA and RNA-RNA ISH with optimized IGSS and silver acetate AMG provides a valuable alternative to other detection systems for ISH. It is sensitive, fast, reliable and sometimes less costly than other methods and gives an improved resolution.

ACKNOWLEDGMENTS

Results demonstrated in this manuscript were obtained in joint collaboration with Prof. Gorm Danscher (Aarhus, Denmark), Dr. Robyn Rufner (Browns Mills, NJ, USA), Dr. Wolfgang H. Muss (Salzburg, Austria) and Prof. Erik Wilander (Uppsala, Sweden). The work was partly financed by the Swedish Research Council (Project No. 102; Uppsala, Sweden) and by a research grant from the Deborah Research Institute (Browns Mills, NJ, USA).

REFERENCES

1. **Ballard, S.G. and D.C. Ward.** 1993. Fluorescence *in situ* hybridization using digital imaging microscopy. J. Histochem. Cytochem. *41*:1755-1759.
2. **Bartsch, O. and E. Schwinger.** 1993. A simplified protocol for fluorescence *in situ* hybridization with repetitive DNA probes and its use in clinical cytogenetics. Clin. Genet. *40*:47-56.
3. **Bauman, J.G.J.** 1985. Fluorescence microscopical hybridocytochemistry. Acta Histochem. Suppl. XXXI:9-18.
4. **Bloch, B., R.J. Milner, A. Baird, U. Gubler, C. Reymond, P. Bohlen, D. Le Guellec and F.E. Bloom.** 1984. Detection of the messenger RNA coding for preproenkephalin A in bovine adrenal by in situ hybridization. Regul. Pept. *8*:345-354.
5. **Bloch, B.** 1993. Biotinylated probes for *in situ* hybridization histochemistry: Use for mRNA detection. J. Histochem. Cytochem. *41*:1751-1754.

6. **Breitschopf, H., G. Suchanek, R.M. Gould, D.R. Colman and H. Lassmann.** 1992. *In situ* hybridization with digoxigenin-labeled probes: Sensitive and reliable detection method applied to myelinating rat brain. Acta Neuropathol. *84*:581-587.
7. **Brigati, D.J., D. Myerson, J.J. Leary, B. Spalholz, S.Z. Travis, C.K.Y. Fong, G.D. Hsiung and D.C. Ward.** 1983. Detection of viral genomes in cultured cells and paraffin embedded tissue sections using biotin-labelled hybridization probes. Virology *126*:32-50.
8. **Burns, J., A.K. Graham, C. Frank, K.A. Fleming, M.F. Evans and J.O'D. McGee.** 1987. Detection of low copy human papilloma virus DNA and mRNA in routine paraffin sections of cervix by non-isotopic in situ hybridisation. J. Clin. Pathol. *40*:858-864.
9. **Chan, V.T.-W. and J.O'D. McGee.** 1990. Non-radioactive probes: Preparation, characterization, and detection, p. 59–70. *In* J.M. Polak and J.O'D. McGee (Eds.), *In Situ* Hybridization. Principles and Practice. Oxford University Press, Oxford, UK.
10. **Coates, P.J., P.A. Hall, M.G. Butler and A.J. D'Ardenne.** 1987. Rapid technique of DNA-DNA *in situ* hybridisation on formalin fixed tissue sections using microwave irradiation. J. Clin. Pathol. *40*:865-869.
11. **Coggi, G., P. Dell'Orto and G. Viale.** 1986. Avidin-biotin methods, p. 54–70. *In* J.M. Polak and S. Van Noorden (Eds.), Immunocytochemistry - Modern Methods and Applications. Wright, Bristol, UK.
12. **Coghlan, J.P., J.D. Penschow, P.J. Hudson and H.D. Niall.** 1984. Hybridization histochemistry: use of recombinant DNA for tissue localizations of specific mRNA populations. Clin. Exp. Hypertens. [A] *6*:63-78.
13. **Danscher, G.** 1981. Histochemical demonstration of heavy metals. A revised version of the sulphide silver method suitable for both light and electron microscopy. Histochemistry *71*:1-16.
14. **Danscher, G.** 1981. Localization of gold in biological tissue. A photochemical method for light and electronmicroscopy. Histochemistry *71*:81-88.
15. **Danscher, G. and J.O. Norgaard.** 1983. Light microscopic visualization of colloidal gold on resin-embedded tissue. J. Histochem. Cytochem. *31*:1394-1398.
16. **Danscher, G.** 1984. Autometallography. A new technique for light and electron microscopical visualization of metals in biological tissue (gold, silver, metal sulphides and metal selenides). Histochemistry *81*:331-335.
17. **Danscher, G., G.W. Hacker, J.O. Norgaard and L. Grimelius.** 1993. Autometallographic silver amplification of colloidal gold. J. Histotechnol. *16*:201-207.
18. **Faulk, W. and G. Taylor.** 1971. An immunocolloid method for the electron microscope. Immunochemistry *8*:1081-1083.
19. **Giaid, A., Q. Hamid, C. Adams, D.R. Springall, G. Terenghi and J.M. Polak.** 1989. Non isotopic RNA probes. Comparison between different labels and detection systems. Histochemistry *93*:191-196.
20. **Gu, J., R.I. Linnoila, N.L. Saibel, A.F. Gazdar, B.J. Minna Brooks, G.F. Hollis and I.R. Kirsch.** 1988. A study of myc-related gene expression in small cell lung cancer by *in situ* hybridization. Am. J. Pathol. *132*:13-17.
21. **Gu, J., N. Kulatilake, L. Gonzalez-Lavin, M. D'Andrea and S. Cull.** 1989. Atrial natriuretic peptide and its mRNA in overloaded and overload released ventricles of rat. Endocrinology *125*:2066-2074.
22. **Hacker, G.W., D.R. Springall, S. Van Noorden, A.E. Bishop, L. Grimelius and J.M. Polak.** 1985. The immunogold-silver staining method. A powerful tool in histopathology. Virchows Arch. [A] *406*:449-461.
23. **Hacker, G.W., L. Grimelius, G. Danscher, G. Bernatzky, W. Muss, H. Adam and J. Thurner.** 1988. Silver acetate autometallography: An alternative enhancement technique for immunogold-silver staining (IGSS) and silver amplification of gold, silver, mercury and zinc in tissues. J. Histotechnol. *11*:213-221.
24. **Hacker, G.W., A.-H. Graf, C. Hauser-Kronberger, G. Wirnsberger, A. Schiechl, G. Bernatzky, U. Wittauer, H. Su, H. Adam, J. Thurner, G. Danscher and L. Grimelius.** 1993. Application of silver acetate autometallography and gold-silver staining methods for *in situ* DNA hybridization. Chin. Med. J. *106*:83-92.
25. **Hacker, G.W., I. Zehbe, C. Hauser-Kronberger, J. Gu., A.-H. Graf and O. Dietze.** 1994. *In situ* detection of DNA and mRNA sequences by immunogold-silver staining (IGSS). Cell Vision *1*:30-37.

26. **Hacker, G.W., C. Hauser-Kronberger, A.-H. Graf, G. Danscher, J. Gu and L. Grimelius.** 1994. Immunogold-silver staining (IGSS) for detection of antigenic sites and DNA sequences, p. 19-36. *In* J. Gu and G.W. Hacker (Eds.), Modern Analytical Methods in Histology. Plenum Press, New York.

27. **Hacker, G.W., G. Danscher, L. Grimelius, C. Hauser-Kronberger, W.H. Muss, A. Schiechl, J. Gu and O. Dietze.** 1994. Silver staining techniques, with special reference to the use of different silver salts in light and electron microscopical immunogold-silver staining. *In* M.A. Hayat (Ed.), Immunogold-Silver Staining: Methods and Applications. CRC Press, Boca Raton, FL (In press).

28. **Henke, R.-P., I. Guerin-Reverchon, K. Milde-Langosch, H. Strömme-Koppang and T. Löning.** 1989. *In situ* detection of human papillomavirus types 13 and 32 in focal epithelial hyperplasia of the oral mucosa. J. Oral. Pathol. Med. *18*:419-421.

29. **Henke, R.-P., A. Claviez, T. Löning, G. Rutter and K. Milde-Langosch.** 1989. Immunogold-Silber-Technik zum Nachweis zellulärer Nukleinsäuren und Antigene in Plattenepithelkarzinomen. Acta Histochem. Suppl. *37*:73-78.

30. **Herrington, C.S. and J.O'D. McGee (Eds.).** 1992. Diagnostic Molecular Pathology. A Practical Approach. Vol I and II. IRL Press at Oxford University Press, Oxford, UK.

31. **Höfler, H.** 1990. Principles of *in situ* hybridization, p. 15–29. *In* J.M. Polak and J.O'D. McGee (Eds.), *In situ* Hybridization. Principles and Practice. Oxford University Press, Oxford, UK.

32. **Holgate, C.S., P. Jackson, P.N. Cowen and C.C. Bird.** 1983. Immunogold-silver staining: New method of immunostaining with enhanced sensitivity. J. Histochem. Cytochem. *31*:938-994.

33. **Hudson, P., J. Penschow, J. Shine, G. Ryan, H. Niall and J. Coghlan.** 1981. Hybridization histochemistry. Use of recombinant DNA as a "homing probe" for tissue localization of specific mRNA populations. Endocrinology *108*:353-356.

34. **Kessler, C. (Ed.).** 1992. Nonradioactive Labelling and Detection of Biomolecules. Springer-Verlag, Berlin, Germany.

35. **Krenács, T. and L. Dux.** 1994. Silver-enhanced immunogold-labeling of calcium-ATPase in sarcoplasmic reticulum of skeletal muscle. (Letter) J. Histochem. Cytochem. *42*:967-968.

36. **Lackie, P.M., R.J. Hennessy, G.W. Hacker and J.M. Polak.** 1985. Investigation of immunogold-silver staining by electron microscopy. Histochemistry *83*:545-550.

37. **Lawrence, J.B. and R.H. Singer.** 1985. Quantitative analysis of *in situ* hybridization methods for the detection of actin gene expression. Nucleic Acids Res. *13*:1777-1799.

38. **Lebo, R.V., E.D. Lynch, J. Wiegant, K. Moore, M. Trounstine and M. van der Ploeg.** 1991. Multicolor fluorescence *in situ* hybridization and pulsed field electrophoresis dissect CMT1B gene region. Hum. Genet. *88*:13-20.

39. **Liesi, P., J.-P. Julien, P. Vilja, F. Grosveld and L. Rechardt.** 1986. Specific detection of neuronal cell bodies: *In situ* hybridization with a biotin-labeled neurofilament cDNA probe. J. Histochem. Cytochem. *34*:923-926.

40. **Long, A.A., P. Komminoth, E. Lee and H.J. Wolfe.** 1993. Comparison of indirect and direct *in-situ* polymerase chain reaction in cell preparations and tissue sections. Histochemistry *99*:151-162.

41. **Löning, T. and K. Milde.** 1987. Viral tumor markers, p. 339-365. *In* G. Seifert (Ed.), Morphological Tumor Markers. General Aspects and Diagnostic Relevance. Springer-Verlag, Berlin.

42. **Löning, T., R.-P. Henke, P. Reichart and J. Becker.** 1987. *In situ* hybridization to detect Epstein-Barr virus DNA in oral tissues of HIV-infected patients. Virchows Arch. [A] *412*:127-133.

43. **Maddox, P.H. and D. Jenkins.** 1987. 3-Aminopropyltriethoxysilane (APES): A new advance in section adhesion. J. Clin. Pathol. *40*:1256-1257.

44. **Miller, M.A., P.E. Kolb and M.A. Raskind.** 1993. A method for simultaneous detection of multiple mRNAs using digoxigenin and radioisotopic cRNA probes. J. Histochem. Cytochem. *41*:1741-1750.

45. **Niedobitek, G., T. Finn, H. Herbst and H. Stein.** 1989. *In situ* hybridization using biotinylated probes. An evaluation of different detection systems. Path. Res. Pract. *184*:343-348.

46. **Nuovo, G.J.** 1991. Detection of human papillomavirus DNA in formalin-fixed tissues by *in situ* hybridization after amplification by polymerase chain reaction. Am. J. Pathol. *139*:847-854.

47. **Nuovo, G.J., F. Gallery, P. MacConnell, J. Becker and W. Bloch.** 1991. An improved technique for the *in situ* detection of DNA after polymerase chain reaction amplification. Am. J. Pathol. *139*:1239-1244.

48. **Polak, J.M. and J.O'D. McGee (Eds.).** 1990. *In Situ* Hybridization: Principles and Practice. Oxford University Press, Oxford.

49.**Raap, A.K., F.M. van der Rijke, R.W. Dirks, C.J. Sol, R. Bloom and M. van der Ploeg.** 1991. Bicolor fluorescence *in situ* hybridization to intron and exon mRNA sequences. Exp. Cell Res. *197*:319-322.
50.**Roth, J., P. Saremaslani and C. Zuber.** 1992. Versatility of anti-horseradish peroxidase antibody-gold complexes for cytochemistry and *in situ* hybridization: Preparation and application of soluble complexes with streptavidin-peroxidase conjugates and biotinylated antibodies. Histochemistry *98*:229-236.
51.**Roth, J., P. Saremaslani, M.J. Warhol and P.U. Heitz.** 1992. Improved accuracy in diagnostic immunohistochemistry, lectin histochemistry and *in situ* hybridization using a gold-labeled horseradish peroxidase antibody and silver intensification. Lab. Invest. *67*:263-269.
52.**Schmidbauer, M., H. Budka and P. Ambros.** 1988. Comparison of *in situ* DNA hybridization (ISH) and immunocytochemistry for diagnosis of herpes simplex virus (HSV) encephalitis in tissue. Virchows Arch. [Pathol. Anat.] *414*:39-43.
53.**Springall, D., G.W. Hacker, L. Grimelius and J.M. Polak.** 1984. The potential of the immuno-gold-silver staining method for paraffin sections. Histochemistry *81*:603-608.
54.**Stierhof, Y.-D., B.M. Humbel, R. Hermann, M.T. Oten and H. Schwarz.** 1992. Direct visualization and silver enhancement of ultra-small antibody-bound gold particles on immunolabeled ultrathin resin sections. Scanning Microsc. *6*:1009-1022.
55.**Varndell, I.M., J.M. Polak, K.L. Sikri, C.D. Minth, S.R. Bloom and J.E. Dixon.** 1984. Visualisation of messenger RNA directing peptide synthesis by *in situ* hybridisation using a novel single-stranded cDNA probe. Histochemistry *81*:597-601.
56.**Zehbe, I., E. Rylander, A. Strand and E. Wilander.** 1992. *In situ* hybridization for the detection of human papillomavirus (HPV) in gynaecological biopsies. A study of two commercial kits. Anticancer Res. *12*:1383-1388.
57.**Zehbe, I., G.W. Hacker, J. Sällström, E. Rylander and E. Wilander.** 1992. *In situ* polymerase chain reaction (*in situ* PCR) combined with immunoperoxidase staining and immunogold-silver staining (IGSS) techniques. Detection of single copies of HPV in SiHa cells. Anticancer Res. *12*:2165-2168.
58.**Zehbe, I., G.W. Hacker, W.H. Muss, J. Sällström, E. Rylander, A.-H. Graf, H. Prömer and E. Wilander.** 1993. An improved protocol of *in situ* polymerase chain reaction (PCR) for the detection of human papillomavirus (HPV). J. Cancer Res. Clin. Oncol. Suppl. *119*:22.
59.**Zehbe, I., E. Rylander, A. Strand and E. Wilander.** 1993. Use of probemix and omniprobe biotinylated cDNA probes for detecting HPV infection in biopsy specimens from the genital tract. J. Clin. Pathol. *46*:437-440.
60.**Zehbe, I., J. Sällström, G.W. Hacker, E. Rylander, A. Strand, A.-H. Graf and E. Wilander.** 1994. Polymerase chain reaction (PCR) *in situ* hybridization: Detection of human papillomavirus (HPV) DNA in SiHa cell monolayers, p. 297-306. *In* J. Gu and G.W. Hacker (Eds.), Modern Analytical Methods in Histology. Plenum Press, New York.
61.**Zehbe, I., G.W. Hacker, J.F. Sällström, E. Rylander and E. Wilander.** 1994 Self-sustained sequence replication-based amplification for the *in situ* detection of mRNA in cultured cells. Cell Vision *1*:20-24.

Address correspondence to Gerhard W. Hacker, Salzburg General Hospital, Institute of Pathological Anatomy, Immunohistochemistry and Biochemistry Unit, Muellner Hauptstr. 48, A-5020 Salzburg, Austria.

In Situ PCR: New Frontier for Histopathologists

Virginia M. Anderson

Department of Pathology, State University of New York, Health Science Center of Brooklyn, Brooklyn, NY, USA

BACKGROUND

In 1993, Nobel laureate Kary Mullis was honored for the discovery of the polymerase chain reaction (PCR) (5). This highly sensitive technique is capable of amplifying a single gene copy to a level that can be readily detected by gel electrophoresis and confirmed by Southern blot analysis. Conventional PCR assays destroy the cells or tissue source in order to extract DNA. The standard *in situ* hybridization (ISH) technique preserves morphology but is relatively insensitive as compared to PCR (4). Well-established protocols have been developed during the past decade, and ISH has become an essential tool in pathology laboratories for the tissue diagnosis of viruses and tumors. Target DNA or RNA is hybridized with a labeled nucleotide probe, and the cellular tissue reaction can be seen under the light or electron microscope.

The inevitable marriage of PCR and ISH merges two highly successful new technologies into a remarkably sensitive method that can detect minute reaction products while preserving tissue morphology. Difficulty in transferring the PCR procedure from a tube to a tissue slide has imposed a formidable challenge.

ESSENTIAL ELEMENTS

PCR and *in situ* PCR (ISPCR) require the application of primers and nucleotides to a test sample that may be either in solution in a test tube or in a cell or tissue sample on a glass slide. A heat-resistant enzyme, *Taq* DNA polymerase and an automatic thermocycler with a stable heat block to control precise temperature differentials are required for the following sequence of reactions to proceed. Thirty 0.5- to 1-min heat cycles of denaturing at about 95°C, of annealing at about 65°C and of extension at about 72°C, may increase the sample size by one billionfold in the PCR mixture. Heat coagulates protein and the outlines of histologic landmarks may be blurred when ISPCR is performed on tissue sections. Experience in routine histology is a must for accurate morphological analysis. Tissue on silanized slides was originally placed in contrived oil-filled aluminum boats and placed on top of a thermocycler designed for conventional PCR. Recently, models designed specifically for glass slides have become available and are able to accommodate at least 16 slides. Simultaneously analyzed controls are critical for interpretation.

Reagents include enzymes, primers and stable PCR mixtures. Adequate washing procedures must be scrupulously followed. An assay can take up to three days to complete. Positive and negative controls, omission of primers and pretreatment of either DNase or RNase must be evaluated before proper interpretation can be rendered. The extreme sensitivity of PCR makes contamination with extraneous DNA, RNA or their enzymes a serious concern. Tissue preparation and gels or slides should be prepared in two physically separate rooms. Maverick polymerase reactions are costly in consuming reagents and technical time when results are uninterpretable. Another major frustration is the lack of standard protocols to be accepted in multiple laboratories. Confirmatory experiments that are the backbone of viral isolation and identification are difficult when each scientist has a customized procedure. The multiplicity of methods is a clear indication that the best protocol has not yet arrived. Frequent efforts to enhance sensitivity, specificity and reproducibility are under constant revision by the masters of ISPCR.

ISPCR

ISPCR can be designed to detect alternations in genomic DNA. The most critical step involves the determination of the proper temperature in degrees, duration and cycle number. This must be optimized for each specific DNA or RNA of interest. DNA must be denatured, amplicons produced and detection performed with a label such as biotin, digoxigenin, fluorescence, radioisotopes, Nitroblue Tetrazoline (NBT), etc. A single gene copy of DNA is all that is required to provide substrate for PCR. Cross linkages formed during formalin fixation are not an absolute barrier to producing a nucleotide substrate. Cell suspensions, cytospins, smears and archival paraffin-embedded material may be used (4). Results are best when tissue is put into paraffin within 2–3 days after formalin fixation. Storage of tissue in 70% or 95% alcohol may enhance and preserve the signal. ISPCR of intact cells in suspension, cytospin or smear uses the cell or nuclear membrane as a test tube to contain the labeled reaction product. To deliver reagents to the target, minute membrane defects must be produced by proteinase K treatment. Negative cells may be truly negative or, for technical reasons, falsely negative and escape detection. Reservations regarding a negative result can be addressed by using mixing experiments as a control. A 1:10 ratio of HIV-infected lymphocytes is mixed with uninfected cells and the efficacy of PCR can be quantified. These experiments, performed by Dr. Omar Bagasara, have been published in *Cell Vision—Journal of Analytical Morphology*, which has published many advances in the *in situ* PCR field.

Reverse Transcription (RT)-ISPCR

An additional step is required for the identification of RNA. The enzyme reverse transcriptase produces complementary DNA (cDNA), which can then be amplified. Pretreatment with DNase is necessary to eliminate native DNA so as to ensure that the reaction product is based on cloned cDNA only, for this is the true mirror image of the RNA of interest. The application of this procedure is many times more vast than ISPCR, which is based solely on altered genomic DNA. RT-ISPCR permits an analysis of gene expression which results in specific proteins produced by messenger RNA. In the past, emphasis was placed on cellular pathology. Now dynamic stroma maintained as extracellular secretions produced under DNA control may be studied. Previously, informational systems, both proximate and remote, were crudely classified as hormones, cyclic ATP-dependent activators, autocrine loops or enzyme dependent reactions, etc. In the *in situ* world, the cell as seen within tissue is defined by its surrounding cells, stroma, adhesion molecules and cytokines, whose functional status is visualized when mRNA is marked as a RT-ISPCR product. The identification of messenger RNA indicates the functional commitment of the cell. Intracellular reaction products may eventually be quantified, nested or multiplexed to test simultaneous intercellular reactions at the nucleotide level. The histopathologist will be able to see exactly where the substrate is in relationship to cells, stroma, cytokines, secretions, microorganisms, etc. Biochemical and molecular biological reactions can now be seen in the context of tissue histology.

Recognition of the mere presence of a gene or gene product is not enough. The strength of the ISPCR method is the capability of seeing markers for gene activation in tissue in relationship to everything else. This identification of gene products brings a new level of sophistication to the study of cell function and the possibility for analysis of cell- to-cell informational interactions within tissues. *In situ* morphological analysis now means that a measurable effect may be studied in the context of the interstitial microenvironment. Surface receptors, cytokine networks, growth factors, interleukins, toxins, adhesion molecules, etc., as well as pathogens may be amplified to detectable levels, labeled, visualized and, in the future, quantified with computer-assisted analysis. Retrofitting of ISH equipment for automated individualized procedures is currently under development.

Now, the possibility of discovering the mechanism of action of a specific protein or lack of an essential protein will enable tissue-specific reactions to reveal why a particular structural or regulator gene is important. The next logical step is to pursue intracellular genetic immunization to correct defects and prevent morbidity by controlling the phenotype at the molecular level. If regulatory genes are known, control of the gene switch may be possible. It may be possible to treat sickle cell anemia by reversing a gene switch for fetal hemoglobin. Gene therapy research is making significant inroads in this direction.

ISPCR, when performed on tissue, cytospins, cytologic smears or fine needle aspirates, must be interpreted by a pathologist familiar with the technique. Pathology chairmen must ensure that this burgeoning field is developed in the anatomic pathology department so that the professional background in histochemistry, immunocytochemistry and routine histopathology can provide a strong anchor in the new world of molecular diagnostics. Assays use costly reagents and are labor-intensive. A scattershot approach, so common in clinical practice today, is both inappropriate and nonproductive. The assay must begin and end with routine morphology. A systematic indication for analysis will be forthcoming when procedures become standardized and inter-laboratory comparisons interpretable. The worst case scenario is a flood of literature with irreproducible results that ultimately detract from the power of ISPCR. A reputation for nonsense data would inhibit growth of this powerful tool and insert a political contaminant into what promises to be a laser sharp technique. Patent protection is also a formidable financial barrier that may have to be reviewed within the context of the public good. An inexperienced morphologist may misinterpret results. Denatured tissue will be difficult to critique in journal articles. RT-ISPCR is not a colorimetric test like a urine dip stick. It requires intense morphological analysis that stretches the limits of light microscopy beyond ultrastructure into the realm of biochemistry, human genetics and molecular biologic identification and specification.

APPLICATIONS: PRESENT AND FUTURE

The identification of low copy latent viruses justifies promotion of ISPCR as the morphologic tool of the millennium. In the late 1980's, HIV was challenged as the cause of AIDS by Duesberg, a prominent virologist (3), because it was impossible to identify a significant viral burden in the peripheral blood of dying AIDS patients. In 1992, Bagasra (1) in his seminal studies in the *New England Journal of Medicine*, demonstrated an unexpectedly high body burden of HIV using ISPCR methods that correlated with the clinical stage of disease and the CD4 counts. Zevallos' application of ISPCR to the human placenta (2) showed that trophoblasts, Hofbauer cells and fetal endothelium are permissive for HIV. Previous reports by Chawandi and Greco with standard ISH showed rare positivity but failed to reveal the incidence or range of infected cell types. HIV infection is not dependent on the presence of the CD4 receptor. Application of ISPCR techniques to other organs of AIDS patients will permit refinement in our understanding of pathogenic mechanisms involved in HIV cardiomyopathy, HIV encephalopathy, HIV enteropathy, HIV nephropathy, etc. The major symptoms of HIV may be the result of cytokine secretions, as seen in the severe wasting syndrome associated with the opportunistic infection, Mycobacterium avium intracellulare. Cachexin, now commonly called tumor necrosis factor (TNF), may be similarly responsible for AIDS dementia. These hypotheses may be tested with RT-ISPCR techniques.

The allocation of grant funds to pathology departments is essential to move this important work forward. To date, emphasis has been placed on HIV assays performed on peripheral blood. Systematic study of human tissue will be highly informative in defining the natural history of the disease and the mechanisms of viral-host interactions. This research is prerequisite to novel vaccine development.

ISPCR methodology has been used to identify Herpes virus, Cytomegalovirus, Ebstein Barr virus, Hepatitis B virus, Hepatitis C virus, human papillomavirus, Visna virus, polio virus, measles and influenza viruses in formalin-fixed and paraffin-embedded tissue (PET). The mere presence of a virus does not confirm that it is an etiologic agent. Study of tissue reactions may indicate the presence of disease, especially when altered genomic DNA is present. Conventional PCR may be as sensitive as viral culture in determining that a virus is trophic for a particular cell type. However, it is an absolute requirement for viral activation to be confirmed by mRNA-based assays, by direct visualization of tissue reactions or a characteristic clinical picture to confirm that an infectious agent is not only present but is the cause of disease. The pathogen may be viewed as a probe that can decipher the immune response using immunocytochemistry for cell marker identification and RT-PCR to determine patterns of cytokine production. This approach should be particularly informative in studying mucosal immunodeficiency and gut opportunistic infections in HIV disease. Novel immunotherapy strategies or immune modulation may emerge with the development of *per os* vaccines.

Not only will diagnostic accuracy improve, but a new classification of infectious disease and neoplasia should be possible. This is particularly important in risk stratification for leukemia, the detection of minimal residual disease, the monitoring of the emergence of new mutations and evolution of resistance to chemotherapy. Leukemia is a heterogeneous group of diseases. Molecular classification may improve prediction of biologic behavior and response to treatment. The identification and quantification of chromosomal translocations in leukemia can be detected by the application of ISPCR to cell suspensions studied by flow cytometery. Multiplex PCR can test for a cluster of common chromosomal errors simultaneously. Similar approaches can monitor the viral burden of HIV and the efficacy of new drugs to reduce viral load. ISPCR with flow cytometery and CD4 receptor analysis should accelerate the evaluation of new antiretroviral agents. The diagnosis and prognosis of pediatric solid tumors should improve. Specific translocations, such as 11:22 translocation in Ewings sarcoma, or oncogenes, such as n-*myc* in neuroblastoma, can correlate with prognosis independent from histopathologic grading systems.

The fine line between criteria for malignancy, cellular atypia, repair and benign proliferative reactions may be impossible to delineate on morphological grounds alone. Tumor markers, cell receptor assays and gene rearrangements may be applied to confirm a diagnosis and demonstrate monoclonality, which morphology can only suggest. Radiation-induced mutations and genetic damage may incite a second tumor years after a cure from a primary malignancy has been eradicated. The molecular events responsible for the emergence of second tumors can be studied. The effects of exposure to teratogens, environmental toxins or ultraviolet light may uncover new insights into DNA repair, aging and predisposition to malignancy. The molecular regulation of programmable cell death may be interpreted or modulated when basic mechanisms are known. Up-regulation of the bcl-2 gene may inhibit cell death and ward off T cell attrition in HIV disease. PCR techniques promise to unmask pathogenic mechanisms, as well as provide an adjunct for tissue diagnosis and prognosis.

Fine needle aspiration (FNA) cytology may be ideally suited to molecular applications. Drs. Bibbo and Bagasra from Thomas Jefferson University in Philadelphia studied a needle aspirate of metastatic adenocarcinoma in the pleura. The primary site was determined by performing an ISPCR application for both surfactant gene products as expected in a lung primary as well as amplification of the estrogen receptor, which confirms a primary breast carcinoma (5). It will become increasingly possible to make a specific biochemical and morphological diagnosis on smaller tissue fragments or aspirates. A practical low invasive approach to the epidemic of breast cancer could exploit tumor markers that may be followed with a combination of FNA and PCR. Understandably, anxious women may be monitored in a safe, expeditious manner on an outpatient basis. Less invasive procedures have reduced complication rates and health care costs. Also, less threatening procedures may prompt women to seek early intervention and improve long-term outcome.

Genetic predisposition to tumors will be studied from premalignancy to frank malignancy. The identification of the two genes for tuberous sclerosis should allow PCR methodology to determine whether the disease is familial or a new mutation. It may be possible to explain why some patients get cardiac rhabdomyomas and others get central nervous system gliomas or angiomyolipomas of the kidney. The gene exists in germ line DNA, but unknown factors permit tuberin expression in multiple ways.

PCR performed on chromosomes can identify trisomy, translocation, mutations, somatic mosaicism and co-genes. Pedigree analysis can be performed on paraffin blocks. The placenta may reveal fetal disease in suspect cases when autopsy permission was denied. Gene dosage effects and the influence of suppressor genes should be better understood. Lesions of intracellular filaments, such as cyclin or beta actin, may be discovered. With immunogold labeling methods, corresponding ultrastructure may also be performed on the histopathology samples.

The exposure of the microenvironment in normal and disease states must provide new insights into the fundamental mechanism of inflammation. For example, in inflammatory bowel disease, events in the evolution of the hyperplasia-adenoma-carcinoma sequence, currently managed by the unsatisfactory but lifesaving prophylactic colectomy in teenagers with ulcerative colitis, may yield to treatment with biological modifiers when the disease process is better understood. The mechanisms whereby the normal mucosal barrier resists invasion by streams of microorganisms is unknown. Cell surface molecules and adhesion molecules that spill into the microenvironment of the stroma may some day be visualized with RT-ISPCR technology. Alternations in lamina propria, cell populations and cytokines may be mapped and functionally analyzed with ISPCR.

Gene therapy must be monitored to confirm engraftment and proliferation of altered cells. Intracellular immunization as an anti-HIV strategy must be monitored to determine if the donor gene that up-regulates or suppresses the immune system can be activated and replicated in daughter cells. The tools to test empirical approaches bring a scientific method into the clinic to evaluate novel treatment efforts. Bone marrow transplantation for advanced cancer patients will require extensive molecular study. Recently, host chimerism was proven by the discovery of solid organ donor genes in host cells. The program for a gene therapy conference includes discussion of diverse topics, such as the use of enzyme therapy in AIDS to render stem cells resistant to HIV. Gene-modified synovial cells delivered to the joint antagonizes interleukin I receptors to treat crippling rheumatoid arthritis. The functional domain of the dystrophin gene can be transferred to skeletal muscle by adenovirus vectors. Cationic lipid-mediated gene transfer may optimize gene delivery to airway cells in patients with cystic fibrosis. Angiogenic growth factor genes or inhibition of proliferative vascular myocytes in vascular disease may be used to improve blood flow in peripheral vascular disease or prevent re-stenosis in post-cardiac bypass surgery. These approaches are in an early experimental stage.

Gene therapy must be monitored by ISPCR to verify that clinical improvement is not from a placebo effect, but the molecular basis can be proved *in vivo* and *in vitro*. The pathologist of the future must be prepared to support these clinical research activities. It takes years of experience to develop morphologic skills. Unless pathologists go to the bench, the void in ISPCR analysis will be filled with the undisciplined observations of desperate researchers who know not what they see.

Molecular diagnostics promise a new dawn of achievement for anatomic pathology laboratories. New technologies will spawn a new taxonomy and inspire novel treatment that must be monitored. The anatomic pathologists are custodians of the tissue. The tools for understanding hidden biologic truths in tissue will undergo intense development. The challenge is to go beyond the hematoxylin and eosin section and put new and heretofore unanswerable questions to nature.

REFERENCES

1. **Bagasra, O., S.P. Hauptman, H.W. Lischner, M. Sachs and R.J. Pomerance.** 1992. Detection of human immunodeficiency virus type 1 provirus in mononuclear cells by *in situ* polymerase chain reaction. Journal N. Engl. J. Med. 326:1385-1389.
2. **Bibbo, M., J.P. Pestaner, L.M. Scavo, L. Bobroski, T. Seshamma and O. Bagasra.** 1994. Surfactant protein A mRNA expression utilizing the reverse transcription *in situ* PCR for metastatic adenocarcinoma. Cell Vision *1*:290-293.
3. **Duesberg, P.H.** 1991. AIDS epidemiology: Inconsistencies with human immunodeficiency virus and the infectious disease. Proc. Natl. Acad. Sci. USA *88*:1575-1579.
4. **Gu, J.** 1994. Principles and applications of *in situ* PCR. Cell Vision *1*:8-19.
5. **Mullis, K.B.** 1990. The unusual origin of the polymerase chain reaction. Sci. Am. April:56-65.

Address correspondence to Virginia M. Anderson, State University of New York, Health Science Center of Brooklyn, Department of Pathology, Box 25, 450 Clarkson Ave., Brooklyn, NY 11203, USA.

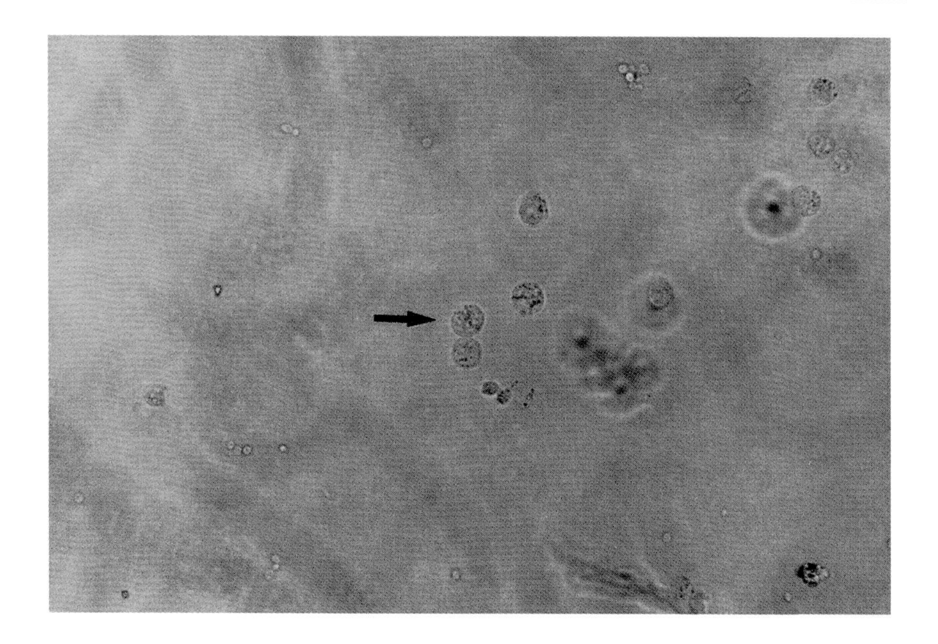

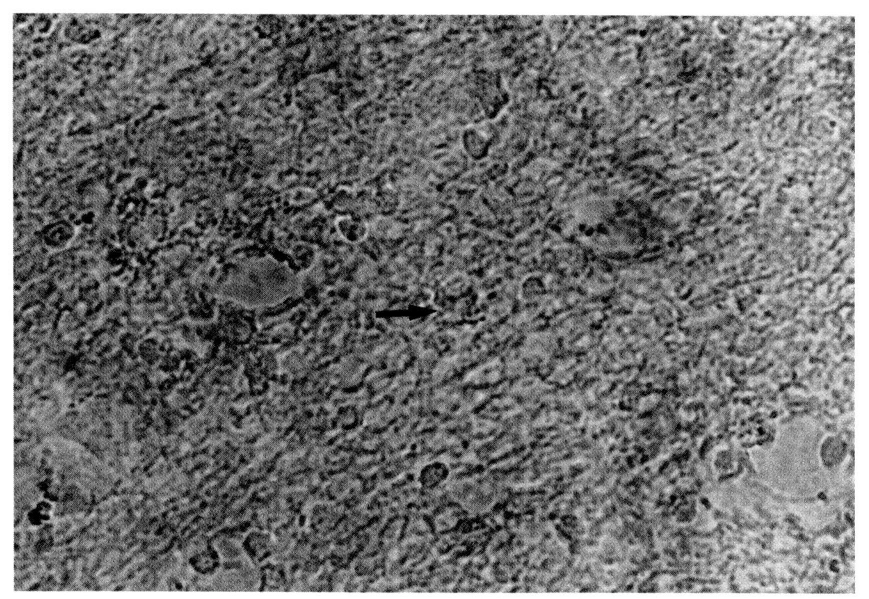

Figure 2. After optimal proteinase K treatment. The so-called salt and pepper dots will begin to appear approximately 5 min after the initiation with proteinase K treatment. The optimal period for the enzymatic treatment is determined when there are around 10–20 "peppery dots" on the cell surface. The best way to visualize the "dots" is to use phase-contrast microscopy. As illustrated above in **A**), there are numerous cells with the peppery dots (arrows). All the cells in this visual field exhibited the presence of the dots. In **B**) a representative section from a brain section is shown. Here the visualization is difficult (arrows) and requires much practice. *(See p. 43)*

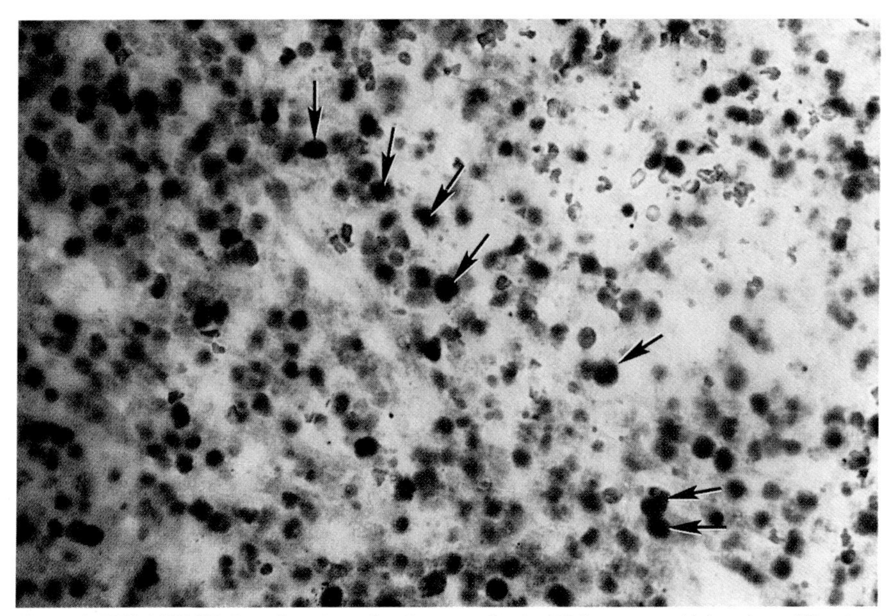

Figure 1. HIV-1 ISPCR. Lymph node section from a pediatric AIDS autopsy with lymphoproliferative syndrome. The *in situ* amplification was accomplished using HIV-1 *gag* primers SK38 and SK39, as described in the HIV ISPCR protocol above. *In situ* hybridization was performed using biotinylated HIV-1 SK19 probe. The majority of lymphocytes show nuclear positivity (blue color). Some positive cells have enlarged nuclei (arrowheads). Nuclear Fast Red counterstain. Magnification, ×20. *(See p. 95)*

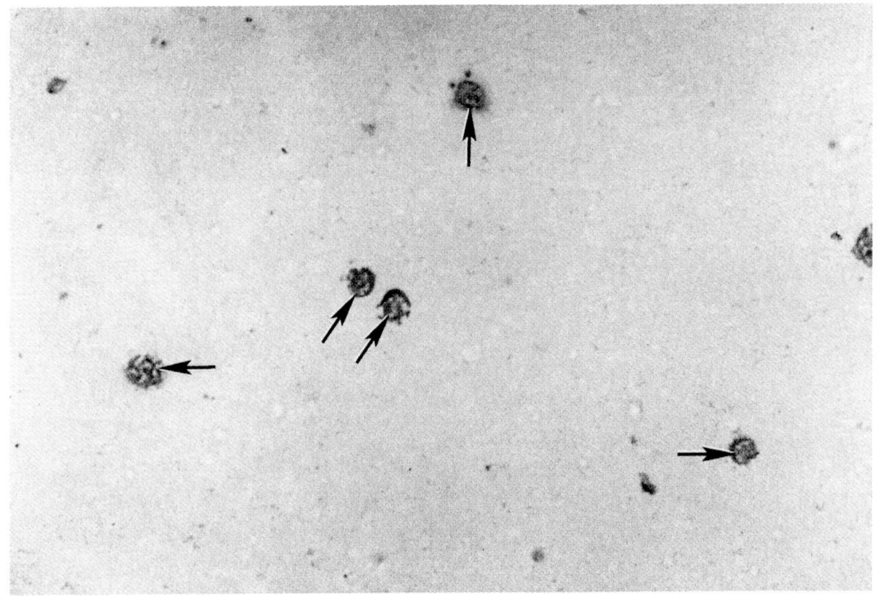

Figure 2. HIV-1 RT-ISPCR. Peripheral blood mononuclear cells from AIDS patient were obtained by Ficoll®-gradient centrifuge and cytospin. The reverse transcription and amplification were achieved using the r*Tth* DNA polymerase Driven RT-ISPCR method with HIV-1 *gag* primers SK38 and SK39. *In situ* hybridization was performed with biotinylated HIV-1 SK19 probe. Lymphocytes (arrowheads) display cytoplasmic positivity (blue color). Nuclear Fast Red counterstain. Magnification, ×40. *(See p. 96)*

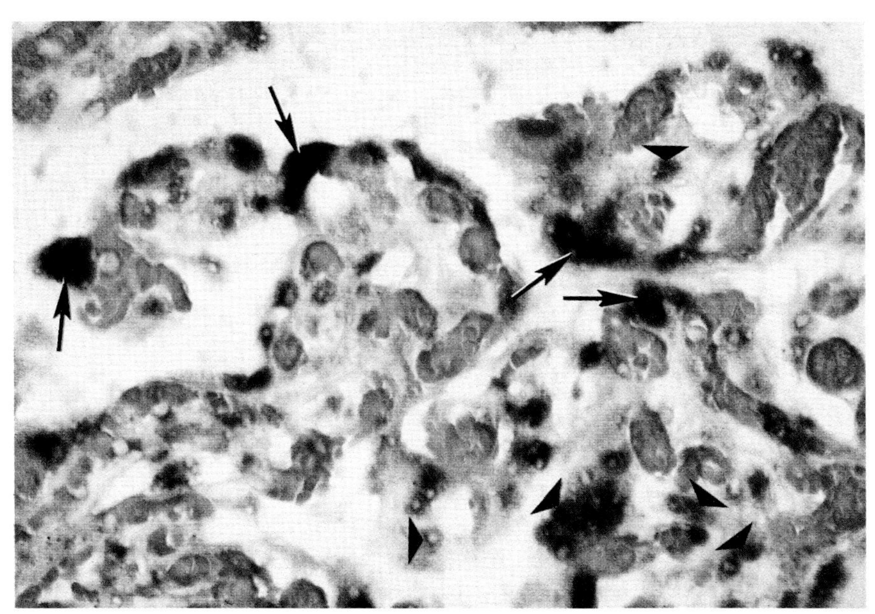

Figure 3. HIV-1 ISPCR. Section of placenta from HIV-infected mother. *In situ* amplification was performed with HIV-1 *gag* primers SK145 and SK431, as described in the HIV ISPCR protocol. *In situ* hybridization was carried out using biotinylated HIV SK102 probe. Syncitotrophoblast (arrows) and Hofbauer cell (arrowheads) display nuclear positivity (blue color). These placental cells are frequently the most positive. Nuclear Fast Red counterstain. Magnification, ×40. *(See p. 96)*

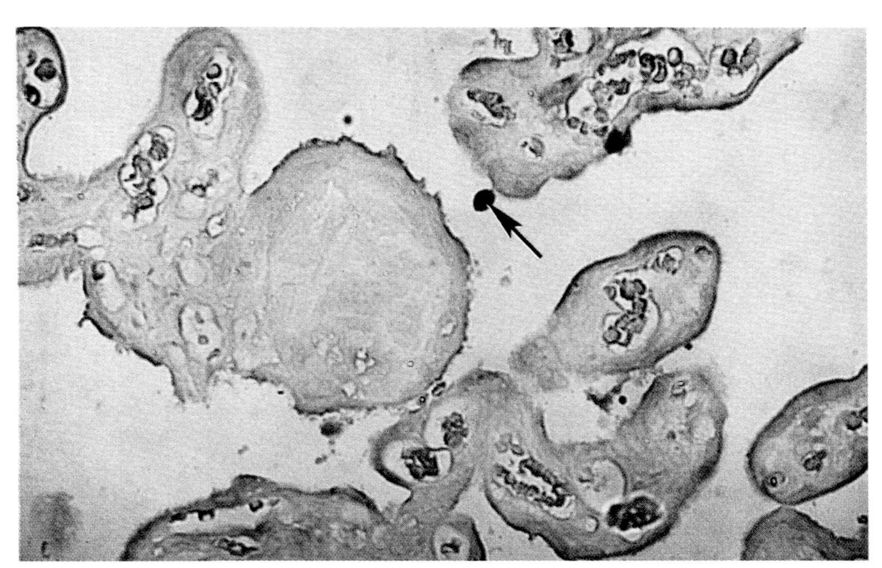

Figure 4. HIV-1 ISPCR. Placental section from HIV-infected mother. *In situ* amplification was performed using HIV-1 *gag* primers SK38 and SK39, and *in situ* hybridization and detection were carried out according to the described HIV ISPCR protocol. The arrowhead points to a positive lymphocyte (dark blue) of the intervillous space in close contact with the trophoblastic layer. Cell-to-cell interaction as demonstrated here may be an important route in the pathogenesis of vertical transmission of HIV-1. Fast Green counterstain. Magnification, ×20. *(See p. 97)*

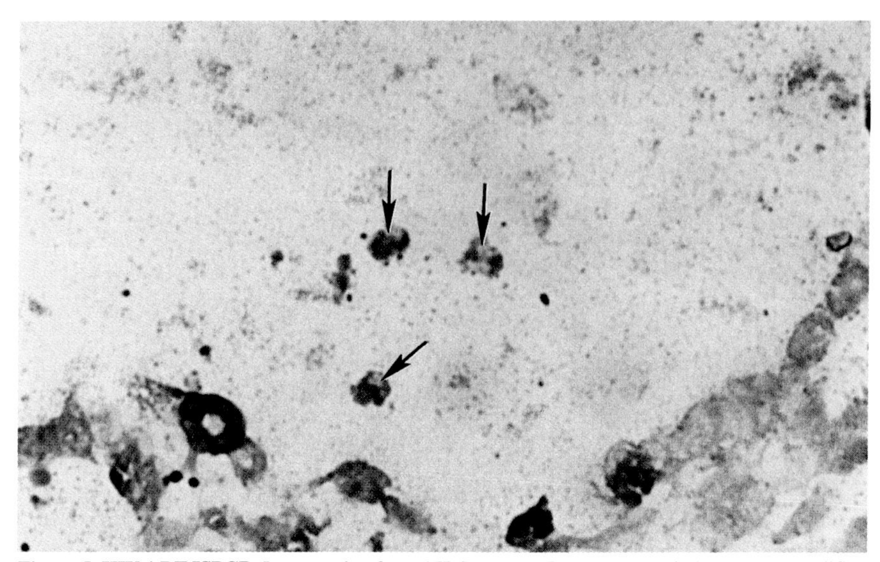

Figure 5. HIV-1 RT-ISPCR. Lung section from AIDS autopsy. Reverse transcription, *in situ* amplification, hybridization and detection were performed according to the Reverse Transcriptase Driven RT-ISPCR protocol. The initial reverse transcription was achieved with random hexamers. The *in situ* amplification was performed by using HIV-1 *gag* primers SK38 and SK39. Arrowheads point to cytoplasmic positivity (blue color) almost exclusively in intra-alveolar cells with macrophage morphology. Although there is significant distortion, some faint positivity is seen in other intra-alveolar cells. The relevance of this finding is that intra-alveolar macrophages may be the sites of viral replication indicating an "active" infection. Nuclear Fast Red counterstain. Magnification, ×40. *(See p. 97)*

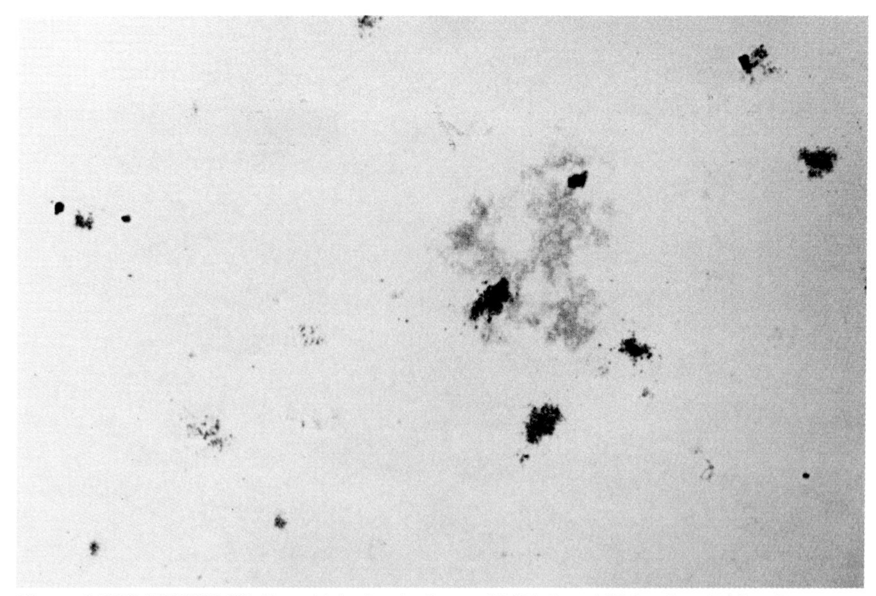

Figure 6. HIV-1 RT-ISPCR. Bronchioloalveolar lavage (BAL) from AIDS patient. BAL cells were cytospun on glass slides. Reverse transcription, *in situ* amplification, hybridization and detection were performed as previously described. Moderate heat distortion is observed. Strong cytoplasmic positivity (dark blue) is present in cells with macrophage morphology. The pink amorphous structure in the center corresponds to epithelial cells. Pyronin Y counterstain. Magnification, ×40. *(See p. 98)*

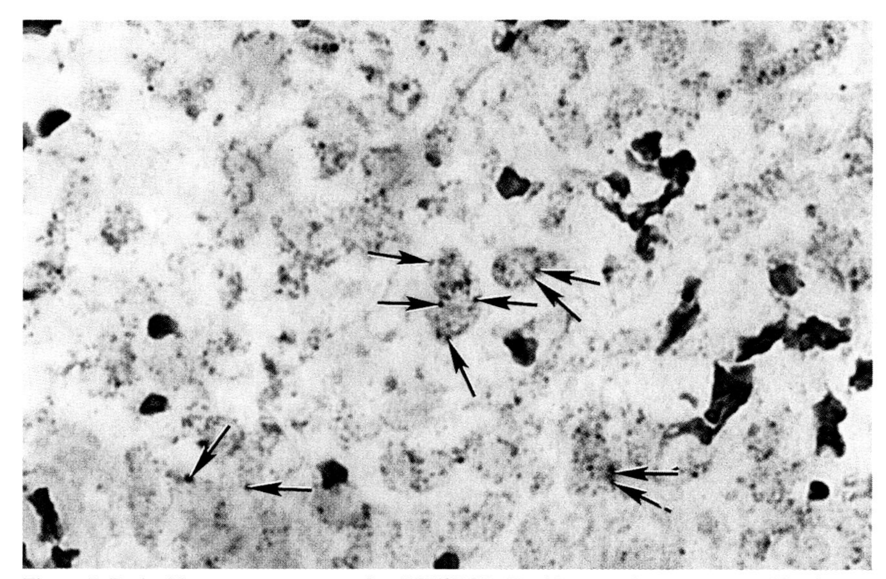

Figure 7. Retinoblastoma gene expression RT-ISPCR. Fetal lung section. *In situ* amplification and detection were performed according to the r*Tth* DNA Polymerase Driven RT-ISPCR protocol. Sections were pretreated with RNase-free DNase overnight. Specific retinoblastoma complementary primers for mRNA. Biotin-dUTP were added to the PCR mixture for direct labeling of amplified products. Granular positivity is restricted to cytoplasmic location (arrowheads) in most of the cells. No nuclear positivity is observed, which indicates absence of "pseudo-primer false positivity" (see text). Direct labeled ISPCR with specific RNA complementary for commonly expressed genes can be useful to verify amplifiability of mRNAs. Magnification, ×40. *(See p. 98)*

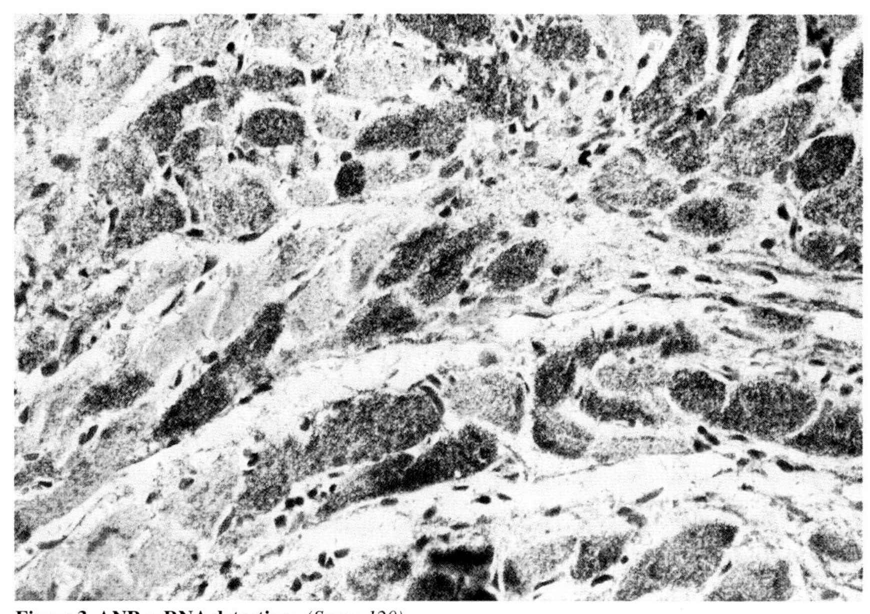

Figure 3. ANP-mRNA detection. *(See p. 120)*